Toys in Space
Exploring Science
with the Astronauts

Toys in Space
Exploring Science with the Astronauts

Dr. Carolyn Sumners
Project Director for the Toys in Space Program

Foreword by John Casper
Commander, Endeavour, STS 54

TAB Books
Division of McGraw-Hill, Inc.
Blue Ridge Summit, PA 17294-0850

FIRST EDITION
FIRST PRINTING

© 1994 by **Carolyn Sumners**.
Published by TAB Books.
TAB Books is a division of McGraw-Hill, Inc.

Library of Congress Cataloging-in-Publication Data
Sumners, Carolyn.
    Toys in space : exploring science with the astronauts / by Carolyn
Sumners.
        p.   cm.
    Includes index.
    ISBN 0-8306-4533-0   ISBN 0-8306-4534-9 (pbk.)
    1. Astronautics—Research. 2. Toys—Study and teaching
(Elementary) 3. Weightlessness—Experiments. 4. Space shuttles-
-Experiments. I. Title.
TL860.T69S86 1993
500.5—dc20   93-21036
                                                                    CIP

Acquisitions Editor: Kimberly Tabor
Editorial team: Laura J. Bader, Editor
            Susan Wahlman Kagey, Managing Editor
            Joanne Slike, Executive Editor
Production team: Katherine G. Brown, Director
            Ollie Harmon, Typesetting
            Bonnie Marie Paulson, Typesetting
            Lori L. White, Proofreading
            Wendy L. Small, Layout
            Nadine McFarland, Quality Control
            Joann Woy, Indexer
Design team: Jaclyn J. Boone, Designer
            Brian Allison, Associate Designer
Art: On-orbit photographs by the National Aeronautics and Space Administration. All other pho-
    tographs by Gary Young, The Vela Group, 3300 Kingswood Lane, Houston, TX 77092
        Enhancements of video images from the STS 54 mission were made from NASA's on-orbit
    videotapes by Gary Young, The Vela Group.
        Illustrations by Rhonda Fleming   AV1
Cover design: Gary Young, The Vela Group, 3300 Kingswood Lane, Houston, Texas 77092   4481

To the men in my life:
My older sons, Rob and Jon, who have always
had to share toys with their mother;
My young son Scott, who arrived nine days before
the liftoff of the Toys in Space I project
and helped in the pilot testing for Toys in Space II;
and my husband, Gary, who brings his own special brand
of sanity to a household overflowing with projects and toys.

In memory of astronaut David Griggs, an integral part of the Toys in Space program, who died in 1989. STS 51D was his only mission.

"I've seen you playing . . . excuse me, I mean demonstrating, with balls and jacks and yo-yos and even a slinky toy in the zero gravity of space. And now I know you're doing this to make some educational videotapes for students learning about the laws of physics. That's really the best thing about our space program: the inspiration and challenge it gives our young people."

President Ronald Reagan,
addressing the astronauts on board Shuttle Flight STS 51D
April 1985

Contents

Foreword

FLYING IN SPACE is fun. We go into space for many serious reasons—to explore, to discover, to deploy satellites, to study the earth or stars, and to conduct scientific and medical experiments, just to name a few reasons. But the overriding truth that becomes obvious to all space flyers once in orbit is that it's enjoyable, an experience quite unlike any on earth. Whether looking at the earth speeding by from 160 miles altitude or floating effortlessly across the crew compartment of the space shuttle, the sense of this unique environment brings out the child in each of us. Like children, we learn how to move around in our new surroundings by floating, tentatively at first, then with more confidence. We watch objects with childlike fascination to discover how they behave in this apparent zero-gravity world. Watching them, we rediscover our sense of wonder.

As astronauts, we are keenly aware of the importance of education and learning. After all, education (along with practical experience) was an important reason we were selected to become astronauts. Once a part of the space program, we continue a never-ending process of learning, rehearsing, reviewing, and relearning the myriad details we must master to be space travelers. Education is not just a concept to us; it's our means for survival.

We also realize the value of dreams. For me and many of my fellow astronauts, flying in space was a childhood dream come true. "How can we stimulate children today," we would ask ourselves, "with the same sense of awe and desire that led us on the path to spaceflight?"

So in our efforts to share our spaceflight experience with others in a way that would be fun, educational, and perhaps even inspirational, the *Endeavour* crew teamed up with Carolyn Sumners. Carolyn masterminded the original Toys in Space project that was conducted on mission STS 51D and continues her fervent efforts to create new ways for children to discover the joy and excitement of science and technology in the world around us.

This book accurately describes how familiar toys behave in the space environment where the downward pull of gravity is absent. It clearly documents those principles of physics that explain why the toys behave as they do, which is often easier to understand in space, without the influence of gravity. But what I like best about this book is the human story that Carolyn has so expertly woven into the fabric of this teaching resource document. For the story of spaceflight is first and foremost a story of the human spirit, the endless quest for knowledge and understanding of this incredible world that our Creator has given us.

John Casper
Commander of *Endeavour*, STS 54

Introduction

TOYS HAVE FOUR unique advantages in bringing science and technology concepts to children:

- Toys address the real world of children. Toys are the tools and technology of childhood. Astronauts have chosen to take toys in space because toys are an integral part of a child's real-world science experience base. Through astronaut toy demonstrations, we are telling children that understanding how things work is important—on earth and in space.

- Toys address the relevance of play. In yesterday's agrarian community, play prepared a child for work. Mastery of the skills developed in childhood on the farm ensured success as an adult. Today's television, video games, and fashion dolls do little to make our children active and effective problem solvers. The thirty-nine toys sent into space give children of all ages the opportunity to experiment and to construct their own knowledge of physics concepts.

- Toys address the interests of every child. Toys speak the language of every child and appeal to the entire student population. Toys are the most exciting and nonthreatening packaging possible for physics lab equipment. Toys appeal to the student who has had limited success in a traditional classroom environment.

- Toys address the needs of every child. As a nation, we are not making science accessible to the masses we have promised to educate, and we are sending the message to a large percentage of the school population that science is something they cannot learn. Toys are universally accessible—not an experience only for the rich, the well-educated, or a specific gender, racial, or ethnic group. All children can learn to play and problem solve with toys. By enhancing the child's world of play, motion toys can make a real contribution toward leveling the playing field on which all children must compete and succeed.

This book describes the adventures of twelve astronauts and thirty-nine toys on two different space shuttle flights. Whenever possible, the story includes the human adventure of spaceflight as well as the in-orbit performance of familiar toys. Physics facts and astronaut feelings combine in this unique vicarious space mission for children. Hopefully readers of all ages will find the story playful, exciting, educational, and fun. Most of the photographs of the astronauts with the toys have been reproduced from the videotape the astronauts shot during their shuttle missions.

USING THE TOYS IN SPACE PROGRAM

This is not a cover-to-cover book, but a resource for designing toy adventures. Read or peruse the first three chapters as your guide for a toys-in-space adventure. Chapter 1 provides the background for the Toys in Space project. Chapter 2 provides the physics concepts you will need to explain toy behaviors on earth and in space. Chapter 3 provides a detailed description of everyday life in orbit and a context for the toy experiments.

Read chapters 4 through 12 in any order you choose, based on the toys you have available or find most interesting. Then group the toys by topic areas, discuss how the toys work on earth, and predict how they will work in space. Many of these chapters also contain instructions for constructing one or more of the toys. Make these toys and then try the different activities.

Chapter 4 describes the adventures of the space acrobats (the flipping mouse, spring jumper, climbing astronaut, and grasshopper). Chapter 5 describes the adventures of the space movers (the swimming frog, fish, submarine, flapping bird, double propeller, and balloon helicopter). Chapter 6 describes the adventures of the space flyers (the spiral plane, ring-wing glider, Bernoulli blower, whirly bird, maple seed, and boomerang). Chapter 7 describes the adventures of the space wheels (the come-back can and the friction car in its track). Chapter 8 describes the adventures of the space magnets (the magnetic marbles, magnetic rings on a rod, and magnetic wheel on a track). Chapter 9 describes the adventures of the vibrating space toys (the Jacob's ladder, paddleball, and coiled spring). Chapter 10 describes the adventures of the spinning space toys (the gyroscope, gravitron, push top, tippy top, rattleback, yo-yo, and spinning book). Chapter 11 describes the adventures of the colliding space toys (the ball and cup, racquetballs and billiard balls, and klackers). Chapter 12 describes the adventures of the space games (basketball,

NASA

Introduction *Introduction* **xv**

jacks, horseshoes, and ball darts). Chapter 13 provides an account of the trip back to earth.

The appendices are for your reference. Appendix A provides a personal history for all twelve of the astronauts who donated their time and talents to the Toys in Space project. Appendix B provides a list of the NASA Teacher Resource Centers nationwide. Use it as a guide in acquiring videotapes of the astronauts demonstrating the toys in space.

1

Liftoff !!

ON APRIL 12, 1985, the space shuttle *Discovery* carried eleven toys into microgravity. The STS 54 mission in January 1993 was the second to take toys into space. For both flights, astronauts also carried along the questions of curious children, teachers, and parents who had suggested toy experiments and predicted possible results. A few dozen toys and a few hours of the astronauts' free time could bring the experience of freefall and an understanding of gravity's pull to students of all ages.

NASA

This toy cargo has given the space shuttle program one more role in extending human access to space. Laboratories to explore the heavens, to experiment in microgravity, and to photograph the earth can ride in the shuttle's cargo bay. From this same cargo bay, astronauts can launch earth-orbiting satellites and space probes destined for other planets, and malfunctioning satellites can be repaired in space or brought back to earth. Eventually astronauts will manufacture delicate drugs and special metal alloys in the neutral buoyancy conditions of space. With the addition of a few pounds of toys, the shuttle has also become a space classroom where astronauts can teach the nation's children about living in space.

Final toy preparations for both flights began over a year before liftoff in a conference room in Building 4 of the Johnson Space Center (JSC) in Houston, Texas. These meetings were scheduled as astronaut training sessions. Coordination for the STS 51D flight in April 1985 was handled by Frank Hughes, who is now in charge of astronaut training at the Johnson Space Center. Dr. Greg Vogt, Paul Boehm, and Steve Lawton coordinated the project for the STS 54 mission.

These meetings focused on final equipment selection. Packaging, testing, instruction writing, and planning of camera angles depended on this selection process. For each mission, I arrived with about fifty toys—selected by children and teachers from around the country. Closed doors kept the curious away as the toys rolled, flipped, slid, and bounced across the conference table. Each had been selected by teachers and students as a toy that might do strange and mysterious things in space, and each did its best to earn a shuttle birth.

Both missions had five NASA astronauts. The earlier STS 51D mission also acquired a U.S. senator and payload specialist before liftoff, but they were not involved in the toy selection process.

The astronauts chose old toy favorites as well as toys with unusual gravity-related motions. A few toys required special astronaut training. The yo-yo became space hardware in the talented hands of astronaut David Griggs. The paddleball's wayward motions required the pilot's touch of Don Williams. A ball and cup remained unclaimed until the STS 54 mission when Commander John Casper volunteered to attempt it in space. The STS 51D commander, Karol Bobko, chose two spinning toys— a metallic push top and a traditional gyroscope. Don McMonagle, pilot of the STS 54 flight, also selected spinning toys—an enclosed gyroscope, called a gravitron, and a spinning rattleback.

Mission Specialist Rhea Seddon had played jacks as a young girl, never dreaming that she would one day redesign this familiar game for floating astronauts in space. Mission Specialist Susan Helms picked horseshoes and accepted the challenge of making a ringer in space. Greg

Harbaugh, soon to be nicknamed "Space" Harbaugh, chose a foam basketball and hoop. Jeff Hoffman of STS 51D and Mario Runco of STS 54 both chose the most mechanical toys—the magnetic wheel on a track and the friction car in its loop.

Both flights chose the same mascot—a 2-inch-tall, pink, mechanical flipping mouse. Its first "trainer," Don Williams, decided to call this mechanical creature "Rat Stuff" as a testimonial to the right stuff that every astronaut was supposed to have for a journey into space.

The astronauts rejected many possible space toys. Some were one-of-a-kind, like Vincent, the walking tennis ball, or a wind-up bear that plays cymbals. Others, like balls with tails and jumping beans, lacked the familiarity and availability of the more traditional toys that were chosen. Large toys, expensive toys, and battery-powered toys were also deemed inappropriate.

Plans were made to videotape each toy performing specific experiments in microgravity. Rehearsals were scheduled in the full-size shuttle trainer to determine camera angles and the positioning of the toys within the very small shuttle middeck area. Before the STS 54 flight, many toys were tested in the KC-135 zero-gravity airplane. As this plane dives, passengers feel over 20 seconds of floating freefall—the same feeling the astronauts have while they are in orbit. Although this is not enough time to document most toy behaviors, it does provide an indication of what the toys will do. This aircraft has been nicknamed the "Vomit Comet" because of the way many passengers feel after several dives and climbs.

The double propeller, tippy top, climbing astronaut, and Bernoulli blower had their only freefall experiences in the KC-135. Although they were taken into space, their demonstrations were lowest priority because of the good quality of the video from the KC-135. The ring-wing glider, grasshopper, and whirlybird were also not demonstrated because the STS 54 crew ran out of time. These three toys were tested on a KC-135 flight in March 1993 to collect the remaining data for the toys project. The KC-135 crew included Charles Gaevert, Charles Boehl, Hyang Lloyd, Linda Billica, and Patricia Lowry. Patricia Lowry is the producer of the Toys in Space video for the Johnson Space Center.

While astronauts trained in deploying the toys, flight-ready versions of the toys began the testing procedures required for securing a place on the shuttle. Toys were first investigated in Houston for obvious problems. The boomerang made of balsa wood was replaced by one made of flight certified card stock because the wood might split and produce dangerous wood splinters floating in the shuttle. The wooden shaft of the double propeller was wrapped in thin silver tape to prevent splintering. A spring jumper with a rounded head replaced one with pointed ears that could cause injury. All toys made exclusively of paper were accepted—provided the paper was flight certified. All paper and ink used in space must be made of materials that will not release gases into the shuttle's air supply.

The ocean-in-a-bottle toy was a favorite of many teachers. Equal portions of oil and water with blue food coloring make an ocean with rolling waves inside the bottle. In microgravity, there would be no force to separate the oil and water, and teachers and students were curious about what the liquids would do. Unfortunately no bottle could be found that would contain the oil. When shaken, as during the experience of liftoff, an oily film formed around the screw top—even when tape was applied. After six months of experimenting, this toy was rejected.

Each toy that survived the Houston testing went to White Sands, New Mexico, for outgassing tests in a vacuum chamber. Toys were monitored for escaping gases to see what potentially toxic fumes they might release into the closed environment of the shuttle. Plastic toys proved to be the worst offenders. Special bags were ordered to minimize contact with the crew's recycled air supply. Toys also had to be checked for possible injury to crew members and damage to shuttle equipment. The crew had to guarantee that all the tiny jacks and marbles would be carefully watched as they floated about and that they would be dutifully returned to their plastic bags. Once flight certified, the toys were packed into NASA flight bags and stowed in a middeck locker to await the launch.

But why take toys into space anyway? Motion toys are well behaved and their motions on earth are easily observed. Without exception, these toys obey the physical laws that fill science textbooks. Regardless of their owner's wishes, toys will always play by the rules of physics. Toys are familiar, friendly, and fun—three adjectives rarely associated with physics lessons. Toys are also subject to gravity's downward pull, which often stops their most interesting behaviors. The astronauts volunteered to perform toy experiments in orbit where gravity's tug would no longer have an effect on toy activities. Toy behaviors on earth and in space could then be compared to show how gravity shapes the motions of toys and of all other moving objects held to the earth's surface.

2

Speaking toy science

SPACE IS THE PERFECT physics lab. Objects obey the laws of motion without crashing to the floor or fighting to overcome friction. This is the environment where physics makes sense. This chapter is devoted to speaking science—to mastery of the foreign language of physics—introduced in the simplest of all environments. Fifteen toy motions cover all of the concepts required to describe toy behaviors on earth and to predict their behaviors in space. A bouncing ball, the simplest of all toys, is used to introduce and define most of these new concepts. Use a real or imaginary ball to follow along with each toy motion.

Explanations are simplified and no algebraic expressions are used. Readers who want more equations in their explanations can turn to introductory high school and college physics textbooks. Do not memorize these concepts like strange vocabulary words. Treat physics like a living foreign language. Try to use the concepts in your thoughts and speech. Look for real-world examples—beginning with toys. As you use the concepts, you can construct your own knowledge of what each concept really means. This knowledge will stay with you long after the toy experiments.

TOY MOTION 1: THE MICROGRAVITY FLOAT

Vocabulary: microgravity, freefall

On television, we see astronauts floating about in the space shuttle. They use Velcro to attach themselves and their equipment to walls to keep things from floating out of reach. They eat carefully to keep their food under control.

What causes this weightless sensation? It's *not* the lack of gravity. The earth's pull on orbiting astronauts is almost as strong as it is on the planet's surface. The same gravity force that keeps you in your chair also keeps the astronauts in orbit. If earth's gravity were turned off, orbiting astronauts would fly off into interplanetary space.

What is the most important difference between you in your chair and an astronaut floating in the shuttle? Compare your speeds. You are sitting still relative to the earth's surface. The astronaut is moving at the incredible speed of 8 kilometers per second (km/sec) (28,000 km/hour or 17,500 miles/hour) and circles the earth each 90 minutes. It is this speed that keeps the astronauts floating in orbit.

The shuttle is falling toward earth just like you would be if your chair suddenly broke. But the shuttle is traveling so fast that the earth curves away from it as it falls. If the shuttle were not continually falling toward earth, its horizontal velocity would send it flying away from our planet. The astronauts are falling around the earth without ever reaching the ground (until they slow their speed down and come in for a landing).

So what does it feel like to fall? Imagine jumping off a high diving board. If you sat on a bathroom scale as you fell toward the water, what would the scale read? Zero, of course. You and the bathroom scale are falling at the same rate. Therefore you do not push on the scale and it reads zero. Skydivers and riders on high roller coaster hills have the same weightless sensation.

This same effect happens in orbit. The shuttle, the astronauts, and all of their equipment are falling at the same rate around the earth. So the astronauts and their food, tools, and toys all seem to float in the cabin. If an astronaut were to sit on a bathroom scale while in orbit, it would read zero just as it would for the falling diver.

The best term to describe the astronaut's condition is *freefall* because that is what the astronaut is doing. *Microgravity* is also used because it is the official NASA term for this freefall experience. This is the environment where toys can illustrate basic physics concepts in action.

TOY MOTION 2: THE GRAVITY DROP

Vocabulary: gravity, force, accelerate

Drop a rubber ball. Notice that it falls toward the earth. The pull of *gravity* between the earth and the ball causes this to happen. All pushes or pulls are called *forces*. Gravity is a force. Think about the ball's speed as it drops, then imagine it bouncing on a hard surface. Think about its speed as it rises back to your hand. Observe how quickly it bounces off the floor and how it seems to stop as it reaches your hand. Gravity causes falling objects to *accelerate*. The pull of gravity also causes climbing objects to lose speed. Rockets must produce an upward force to overcome gravity's pull and carry the shuttle into orbit.

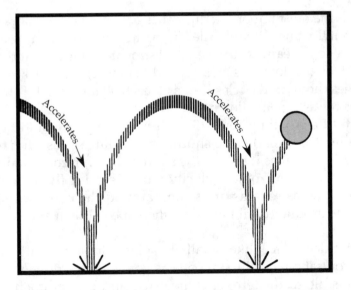

TOY MOTION 3: THE INERTIA ROLL

Vocabulary: inertia, Newton's first law

Push a hard, smooth ball on a hard, smooth, flat, level table. Imagine the ball rolling in the direction that you push it. This straight line motion is caused by the *inertia* principle or *Newton's first law*. According to this law, an object in motion tends to stay in motion unless a force acts on it. A force from your hand starts the ball moving. The ball moves in a straight line because there is no force making it turn. The ball gradually slows down as it rubs against the table and the air. Astronauts in space can expect their toys to move in a straight line at a constant speed.

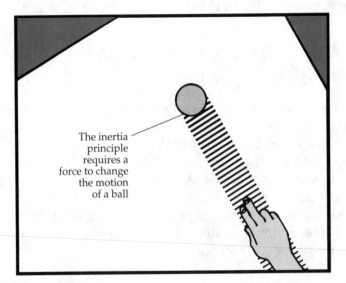

The inertia principle requires a force to change the motion of a ball

TOY MOTION 4: THE FRICTION ROLL

Vocabulary: friction

Roll a ball across a towel. How does its speed change? Friction is the force that slows down the ball. Friction is greater between the ball and the towel than between the ball and a smooth table. Without friction, the moving ball would continue its motion on the flat table. Friction forces occur whenever a moving object rubs against its surroundings. Friction always acts to slow down a moving object.

On earth gravity causes contact between rolling objects and the ground. Friction results from this contact. The rolling motion of wheels and balls minimizes the friction force. Imagine trying to push a ball and a box across the floor. The ball's smaller contact area with the floor lowers the friction and makes the ball easier to push.

Friction is also a friend. Friction provides traction between a basketball shoe and the floor or between a turning car tire and the road. Without this traction, the basketball player would slip and the car would skid out of control. Without friction, astronauts and many of their toys have difficulty moving in space.

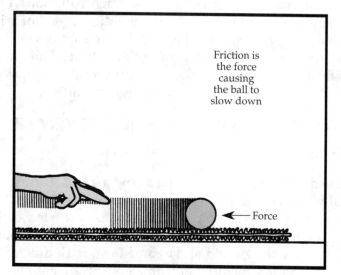

Friction is the force causing the ball to slow down

← Force

TOY MOTION 5: THE MASS EFFECT

Vocabulary: mass, weight, Newton's second law of motion

Imagine two balls of about the same size, but containing different amounts of material—like a small beach ball and a bowling ball. The bowling ball has more *mass* than the air-filled beach ball. The gravity force between the more massive bowling ball and the earth is greater.

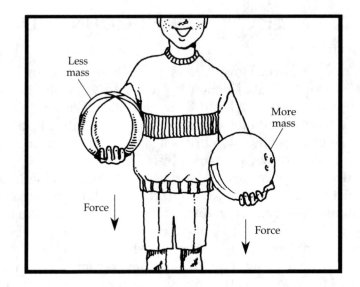

This gravity force is also called the ball's *weight*. The bowling ball weighs more than the beach ball.

Now imagine throwing and catching each ball. Which ball is easier to throw and easier to catch? Obviously the beach ball, but why? As Sir Isaac Newton explains in his second law of motion, more force is needed to start a more massive object moving and then to stop it. This law works in space as it does on earth. Although both the beach ball and bowling ball would float in the space shuttle, astronauts would still have to push harder to start or stop the more massive bowling ball.

TOY MOTION 6: THE GRAVITY RACE

Vocabulary: freefall

Now imagine dropping a beach ball and a bowling ball. The more massive bowling ball feels a greater gravity force as indicated by its greater weight. This greater force is needed to start the more massive bowling ball moving. The famous result is that both balls accelerate toward the ground at the same rate and reach the ground at the same time. Excluding the buoyant effects of the air, all objects, from ants to elephants, fall at the same rate. When an object is falling freely without any support against gravity's pull, it is in *freefall*. Astronauts experience this freefall feeling in orbit.

TOY MOTION 7: THE MOMENTUM ROLL

Vocabulary: momentum

Finally, imagine the beach ball and bowling ball rolling toward you at the same speed. You put out your hands to stop both balls. Which one is harder to stop? Now imagine two identical bowling balls rolling toward you. One is barely moving. The other is barreling across the floor. Which one is harder to stop? All moving objects have *momentum*. A ball that is harder to stop has more momentum. To calculate a ball's momentum, multiply its mass by its velocity. More massive, faster-moving balls have more momentum and are harder to stop.

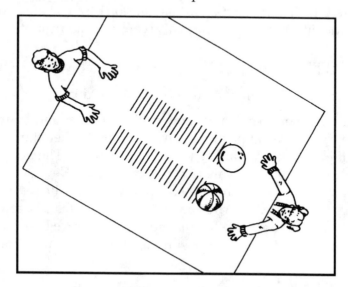

TOY MOTION 8: CONTROLLED COLLISIONS

Vocabulary: conservation of momentum

Roll one marble into another marble of the same mass. Place the marbles in a ruler trough to keep them rolling in a straight line. Notice that the momen-

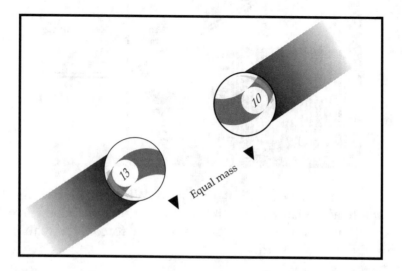

tum of the first marble is transferred to the second marble during the collision. The total amount of momentum is the same before and after the crash. This principle is called the *conservation of momentum*. Pool balls on a billiard table obey the same conservation law. Now roll a more massive shooter marble into a regular marble. Predict what will happen after the collision, then notice how momentum is conserved. Finally, roll a regular marble into a shooter. Predict what will happen in this collision based on momentum conservation. Moving toys in space must also conserve momentum.

TOY MOTION 9: ACTION/REACTION COLLISIONS

Vocabulary: action/reaction forces, Newton's third law

Drop a ball that bounces. Notice how it hits the floor and bounces back toward your hand. What pushes the ball upward? When the ball reaches the floor, it pushes downward. This is the *action force*. The floor reacts by pushing the ball upward. This is the *reaction force*. Newton's third law states that for every action there is an equal and opposite reaction.

Roll a ball into a wall. After impact the ball rolls back toward you. Identify the action force and the reaction force. Notice that the reaction force from the wall is in the opposite direction from the action force of the ball.

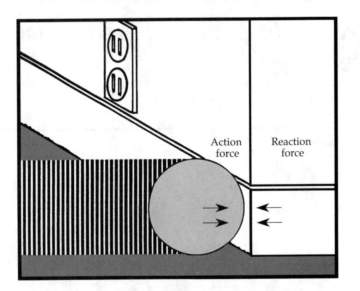

Now think about a car rolling on a road. Which way do its tires push against the road? What pushes the car forward? The action force is the tire pushing backward on the road. The reaction force is the road pushing the tire and the car forward.

Finally, imagine swimming. Wave your hands as if you were pushing against the water. Which way do your hands push the water? What produces the force that pushes you forward in the water? The action force is your hands pushing backward on the water. The reaction force is the water pushing you forward.

TOY MOTION 10: BALL TOSS

Vocabulary: trajectory, parabola, parabolic path

Toss a ball upward toward a friend. Watch the ball's path. This path is called a *trajectory*. Its shape is called a *parabola*. All tossed balls, regardless of path height, trace out parabolic arcs. The parabolic shape is caused by gravity. All basketball, baseball, volleyball, and tennis players adjust their timing to the shape and speed of this parabolic path. In orbit, the trajectory of a tossed ball is a straight line, not a parabola.

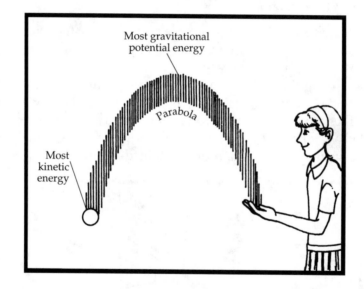

TOY MOTION 11: ENERGY EXCHANGE

Vocabulary: potential energy, kinetic energy, heat energy, conservation of energy

A ball on the edge of a table has *potential energy*. As it falls, its potential energy is converted into the *kinetic energy* of its motion. As the ball bounces and climbs upward again, its kinetic energy turns back into potential energy.

A wound-up toy car has potential energy. As its spring unwinds, the potential energy is converted into the kinetic energy of the moving car.

The stretched elastic of a paddleball has potential energy. When it is released, the potential energy turns into the kinetic energy of the moving ball.

These energy transformations obey a law called the *conservation of energy*. Energy cannot be gained or lost, just transformed from one form to another. Rub your hands together. Notice how your hands feel after a few moments of rubbing. The kinetic energy of your moving hands is converted into *heat energy* by the rubbing.

When a ball rolls down a track, a small amount of its kinetic energy is also converted into heat energy and the ball gradually slows down. Once again energy is conserved. Toys on earth and in space must also conserve energy.

TOY MOTION 12: CIRCLING FORCE

Vocabulary: centripetal force

Attach a string to a ball. In an open area, swing the ball in circles, then release the string. What does your hand feel as you swing the ball? How does the ball move as you release it?

As you swing the ball, you are constantly pulling inward to keep the ball moving in a circle. If you release the string, the ball no longer feels this force and flies away. The inward force that causes the ball to move in a circle is called a *centripetal force*. The ball's motion in a straight line when you release the string is the inertia property talked about in Toy Motion 3.

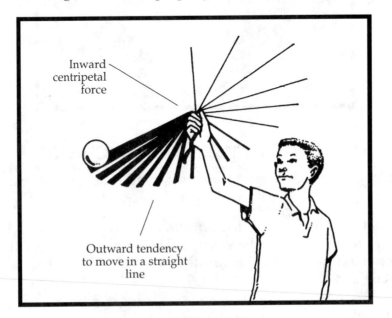

Inward centripetal force

Outward tendency to move in a straight line

Gravity supplies the centripetal force that keeps the shuttle in its orbit. If this gravity force were to vanish, the shuttle would fly off into space.

Note: Sometimes people say that the ball is pulling outward and exerting an outward centrifugal force. The ball seems to be pulling outward because we are forcing it to move in a circle. The author has found that it is better to focus on the centripetal force that actually produces the circular motion and not describe circular motion in terms of any other force.

TOY MOTION 13: SPINNING MOTIONS

Vocabulary: angular momentum, conservation of angular momentum, axis, gyroscopic stability, precession

Put a circular cardboard disk around your pencil and turn it into a top. Spin the pencil and watch its behavior. The pencil becomes the *axis* of your new top.

In Toy Motion 6 you learned that moving objects have momentum. In Toy Motion 8 you saw that this momentum is conserved. In the same way, spinning objects have *angular momentum*, and angular momentum is also conserved. Faster spinning objects have more angular momentum.

Spinning objects also have more angular momentum if their mass is farther from the spin axis. Take the disk off and try spinning your pencil. The result is less angular momentum and a much less stable spin. For this reason, tops and gyroscopes have most of their mass in a wide disk around their spin axis.

A spinning object continues to turn until a force acts on it. On earth, the friction between the top's spinning tip and the table causes the top to slow down. Gravity also tugs on its wide center. The result is a wobbling motion called *precession*.

A gyroscope can spin much faster than a top. You can balance a spinning gyroscope on a string or your finger. Because it conserves angular momentum, a spinning gyroscope is a very stable object. This property is

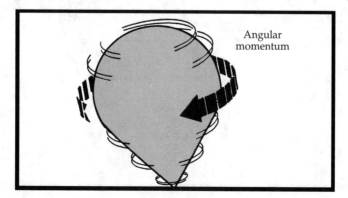

Angular momentum

called *gyroscopic stability*. From spinning satellites to the spinning earth, this gyroscopic stability is a critical component of space flight.

TOY MOTION 14: BALANCE

Vocabulary: center of mass

When a toy spins, it spins around its *center of mass*. Flipping, spinning, and rocking all occur around the center of mass. On earth, acrobats lower their center of mass to remain stable on a balance beam and football players keep their center of mass low during tackles and blocks. Balance is needed to keep from falling on earth. There is no balancing problem in the orbiting shuttle, but the motion of turning toys still relies on the toy's center of mass.

TOY MOTION 15: FLIGHT

Vocabulary: aerodynamic, drag

Toss two pieces of paper, one crumpled into a ball and one that is flat. The ball moves much better through the air because it is more *aerodynamic*. Make a paper airplane and toss it through the air, then try to throw the plane tail first. The ball and airplane going forward are more aerodynamic. The air moves around them instead of stopping their motion. The force of the air that tries to stop a flying object is called *drag*.

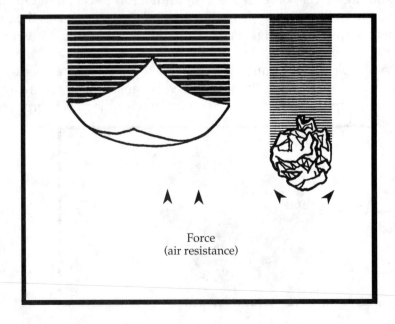

Force
(air resistance)

3

The space environment

THE STS 51D LIFTOFF occurred on April 12, 1985. The STS 54 lifted off on January 13, 1993—almost six years after the *Challenger* tragedy. The first Toys in Space mission was to be a precursor of the lessons taught by the teacher in space, Christa McAuliffe. After the deaths of the seven *Challenger* astronauts, including Christa McAuliffe, the entire space program was put on hold for three years while new safety procedures were developed. Now many shuttle missions have educational objectives like the Toys in Space II project of the STS 54 flight.

ASCENT

Science discoveries begin with the action/reaction of liftoff. Two solid rocket boosters, each carrying over a million pounds of explosives, crackle like giant firecrackers. A combined thrust of over six million pounds carries the spacecraft toward orbit. Expanding engine gases rush from the nozzle, creating a downward action force. The equal and opposite reaction force sends the shuttle skyward. The momentum of the fast-moving gas molecules is matched by the heavy shuttle's upward climb. Using a momentum boost from the earth's rotation, the shuttle soars upward and eastward over the Atlantic Ocean.

Earth-based toys can simulate this liftoff experience. Perhaps the water rocket is the easiest and safest to use and understand. Air fills the rocket's plastic body. A little water is added near the nozzle and is held in place by the pump. As more air is pumped into the rocket, the air compresses. Pressure builds with each stroke, just as pressure builds inside a rocket engine when hot gases try to expand. With the rocket's nose pointed skyward, the pump lever is pulled. Water rushes downward, producing the action force. The lightweight rocket responds by climbing over 50 feet (13 meters). For a higher flight, add more water. The extra

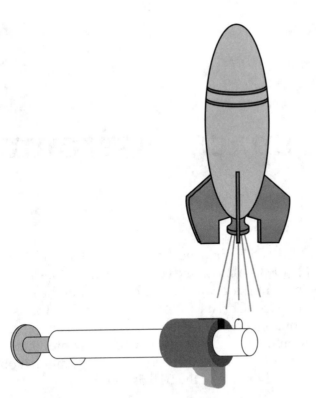

water adds momentum to the escaping water and increases the action force. The rocket responds with a faster liftoff and a higher flight. Of course, a point is reached when additional water cannot escape through the nozzle, and the waterlogged rocket crashes to the ground.

Shuttle launches can generate unpleasant combinations of gravitational and inertial forces. Gravity pulls all passengers and cargo toward earth. The rocket's thrust must overcome gravity while increasing the shuttle's speed. The upward acceleration results in additional forces pushing against the astronaut's back. You can experience a similar inertial force as you sink backward in your seat when a car accelerates rapidly or "burns rubber."

A simple term, the *g*, describes the strength of this acceleration. Sitting comfortably in your chair, reading this book, you are pulling 1 g. The force you feel is your earth weight multiplied by the number 1. In the early days of space travel, an astronaut could expect as many as 8 gs of crushing force at liftoff—the equivalent of seven fellow astronauts piled on top. The shuttle commander can throttle the main orbiter engines to limit the total acceleration to a tolerable 3 gs. An abrupt screeching turn in a car can give a momentary 3-g crush against the car door. A roller coaster pushed upward by a valley curve might deliver over 4 gs of momentary force.

During the shuttle ascent, the full 3-g force comes only twice. Two minutes after liftoff, the two solid rocket boosters drop from the external tank and fall into the Atlantic. Parachutes slow their fall and allow them to be collected and reused. Just before this separation, vertical forces reach the 3-g mark. The second 3-g experience comes 5 minutes later when the shuttle is over 100 miles (160 kilometers) above earth and just before the external fuel tank empties. Once this tank is discarded into the Indian Ocean, the commander controls the last few seconds of climbing and positioning in orbit by firing the orbiter's smaller maneuvering engines.

The journey from the normal 1-g gravity of the earth's surface through the 3-g force of liftoff to the floating weightlessness of orbit takes less than 10 minutes. Astronauts describe this final free-floating feeling as *zero g* or *zero gravity*. NASA calls it *microgravity* since small inertial forces do remain. When no maneuvering engines are in use, any force felt by an orbiting shuttle passenger relative to the shuttle is less than 0.01 g. For moving toys and floating humans, this is effectively freefall. Very sensitive experiments, however, can detect these small forces caused by crew members bouncing off walls and the shuttle's on-board mechanical equipment.

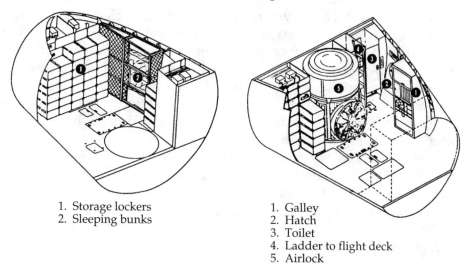

Looking forward:

Looking aft:

1. Storage lockers
2. Sleeping bunks

1. Galley
2. Hatch
3. Toilet
4. Ladder to flight deck
5. Airlock

LIVING QUARTERS

Shuttle astronauts live and work in the cabin's flight deck and middeck areas. Below these decks is a subfloor filled with life-support and maintenance equipment. The flight deck has a two-seat cockpit facing forward and a crew station looking backward into the payload bay. Six windows encircle the forward cockpit. Two face backward into the cargo area and two face upward away from the middeck. The middeck below houses the food galley, sleeping compartments, a restroom, and storage areas. Here experiments, including toy demonstrations, can be performed.

Although the middeck area is too small for large fast-moving toys, it is much roomier than it appears. Sleeping, eating, and playing with toys can occur along the ceiling, floor, or against storage lockers. Without an up or down, astronauts can spread out in all directions, making best use of all available space.

EXERCISE

The astronauts' bodies gradually adjust to this free-floating environment in unusual ways. Blood normally pulled toward the feet by gravity shifts upward. Astronauts' faces fill out and wrinkles disappear. The swelling is sometimes uncomfortable, producing a sensation similar to hanging upside down from monkey bars on a playground. The waistline may shrink 1 to 2 inches (2.5 to 5 centimeters) and pants can become loose. Disks spread apart along the spine and the space fills with liquid. A space traveler gains up to 2 inches (5 centimeters) in height. In the first day of the flight, astronaut Jeff Hoffman grew in height from6 feet, 2 inches to 6 feet, 3.5 inches. Rhea Seddon gained 2 inches to become 5 feet, 3.25 inches tall. On the STS 54 flight, Susan Helms, the youngest crew member, grew 2 inches taller. Commander John Casper, the oldest crew member, gained only 0.5 inch. Space suits must be made to accommodate this astronaut growth spurt. Earth-based humans experience a minor form of this phenomenon each day. Spinal disks relax and spread during the night, resulting in an extra 0.5 inch or so of height in the morning.

In space, relaxed joints assume the midjoint position and the body naturally curves in a gentle arc. When not in use, arms may float out in front of the unsuspecting astronaut. Unused leg muscles begin to atrophy. The heart, with no gravity pull to pump against, slows down. A treadmill with restraining straps and a rowing machine are designed to provide exercise for the heart and lazy leg muscles so that astronauts can stay in shape for the return to earth.

In the first few days of flight, astronauts lose fluid. The heart adapts to a smaller blood volume and the lack of gravity to pump against by growing smaller. Using the American Flight Echocardiograph, Dr. Rhea Seddon measured heart changes during her flight and made pictures of the heart pumping blood.

Many astronauts develop a strange motion sickness called space adaptation syndrome. Nausea, vomiting, and general lethargy can result. Its cause seems related to the absence of any vertical orientation. The brain needs a day or two to adjust to the loss of earthly balance and sense of direction. One of Senator Garn's duties in the STS 51D mission was to monitor his body's reactions to this space sickness.

With three dimensions of floating freedom, simple behaviors become complex tasks. Astronauts quickly miss the firm grip of friction and the presence of a floor. On earth, walking is an action/reaction process made possible by gravity and by friction between the foot and floor. Earth-based travelers push down and back on the floor. As a reaction, the floor pushes them forward and supports them against gravity's tug. The strength of this forward push is predictable.

Space locomotion, in the form of body gliding, takes practice. The novice has trouble deciding how much push to use when leaving a wall and often reaches his destination turned the wrong way and moving so fast that he bounces backward. In his basketball practice, Greg Harbaugh had a similar problem judging his speed and location as he performed several complete 360s before making a basket.

Reaching down to grab something can result in a series of acrobatic somersaults. Rhea Seddon discovered this while trying to collect a way-ward jack. Practice soon teaches that slow motion is mandatory in freefall. If an astronaut's feet slip from their toeholds, he goes limp and waits for his momentum to carry him into something or, occasionally, somebody. Once touched, anything loose will float away. Tasks requiring coordination of many different pieces often take much more time and energy than the frustrated astronaut expects. Keeping plugs, connectors, tools, and the astronaut in the same place long enough to finish a job can become a real chore.

LUNCH IN SPACE

Making a sandwich is an astronaut art requiring crew participation and dedication to task. Nothing rests obediently on the table. Packaged condiments surround the sandwich maker. Blobs of mustard or mayonnaise threaten to float away if shaken or pushed. Bread can drift in air currents if left aloft and unattended.

The perils of flying food hint of the challenges in store for some toy operations. In fact, food itself can become a toy. Don Williams modified his terrestrial juggling skills with a collection of apples and oranges. Weightless juggling has its good and bad points. Dropped fruit will not race to the floor, but tossed fruit will also not return to your hand. The would-be juggler must combine a lateral pitch and catch routine with grab and shift motions. Ever wary of wayward flying fruit, the space juggler must stay alert to positions and velocities. Like many other space motions, the task is tricky. The astronaut's only advantage is that he can practice in slow motion until he gets it right. Fruit can be shifted from place to place at a leisurely pace until toss and grab routines are perfected.

Don Williams is a terrestrial juggler and was eager to lend his talents to the task of space juggling. With cargo weight at a premium, Don Williams selected the most appropriate on-board juggling equipment—apples and oranges. With the help of Commander Bobko, he arranged seven fruits (four apples and three oranges) within his reach. Then he gently moved the fruits from place to place, imitating the behavior of a juggler in slow motion. As simple as it sounds, the task proved challenging. The fruit required constant attention. An apple with a slight wayward drift could be quickly beyond reach. Inertia guaranteed that all fruit would leave the scene following a straight-line path. The juggler's natural tendency to catch an object with an upward push would send the fruit flying away.

With Commander Bobko's help, Don Williams advanced from placement juggling to motion juggling. Several gentle nudges were required to keep each fruit on its box-like circular path. After practicing his space juggling, Don Williams could handle up to four fruits. Unfortunately his juggling fame was short-lived. Each day of practicing saw improvement in his space juggling and a rapid depletion of his props.

FLOATING LIQUIDS

Beverages are consumed through straws. In weightlessness there is no gravity force to keep liquids in their containers. Any drink shaken from a glass floats about the cabin in large blobs. The thirsty astronaut must retrieve his wandering beverage before it splashes into a wall and scatters into a thousand little bubbles. Straws keep liquids under control. Straws must have clamps, however, to stop the drink from flowing once the astronaut finishes taking a sip. No gravity force is available to pull the liquid back into the bottle.

The crew of the STS 51D flight became the first to brew tea in space. The process required pumping hot water from one container to another and back. Both Rhea Seddon and Jeff Hoffman proclaimed the result to be well worth the effort.

Food containers are held in a food tray attached to a table, the astronaut's lap, or the wall. Canned foods and puddings will stay in their opened containers if dining is performed in a slow, deliberate manner.

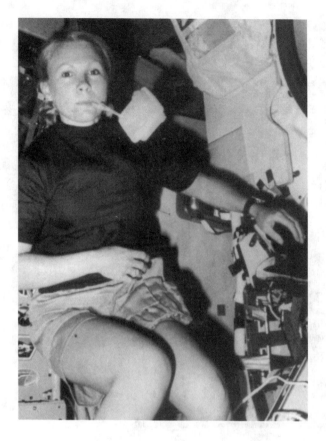

Surface tension can keep the food tray in order until a hard bump or sudden jerk dispenses enough momentum to send food flying across the cabin. Eating utensils behave in earthly fashion, except for spoons which can deliver their cargo at any angle. Food can stick to either side of a spoon. Foods will obediently remain attached to a loose spoon floating through the cabin until retrieved by its hungry owner.

Floating liquids can become the funniest and most fascinating of all toys. A blob of orange juice drifts peacefully through the cabin or hangs playfully in place until a curious astronaut shakes it or blows on it. Even the slightest disturbance produces a shivering throbbing blob that may split into two smaller spheres that drift apart. Blobs will spin with strange currents set in motion around their spin axis. For astrophysicist Jeff Hoffman, each bubble motion made the physics of three-dimensional fluid motions a fascinating observational reality.

Like apples and oranges, bubbles definitely qualify as instant space toys. Unlike earthly soap bubbles, space bubbles are liquid throughout, much like a water balloon without the balloon. They form naturally when

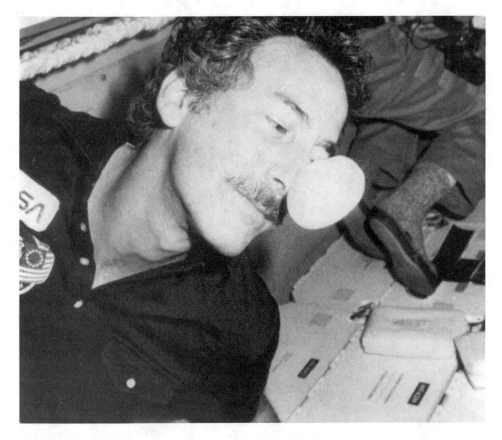

liquids escape from their containers. There is no terrestrial equivalent for such bubbles. Large liquid bubbles cannot float freely in the earth's gaseous atmosphere because the liquid bubble is much heavier than the surrounding air. Gravity always pulls a water balloon to a grassy splashdown.

On earth tiny water droplets form on windows when it rains. They run together and break apart, slipping along the glass. Surface tension holds the water together as gravity pulls each droplet downward. The gravity force limits the size of each tiny round drop. As the drop grows larger, it flattens and finally turns into a flowing liquid. In space, droplets grow into fluid bubbles with no gravity force to pull them apart. The intriguing behaviors of tiny earth droplets hint at the fascinating motions possible in giant floating fluid bubbles. In microgravity, bubbles of water, orange juice, lemonade, or strawberry drink can drift about like apples and oranges until impact. A colliding fruit bounces, but a bubble may shatter into an army of tiny bubbles followed around by a frantic towel-laden astronaut.

Nonspinning bubbles are spherical. Surface tension draws the water into a shape with the same pressure over the surface. This surface is a sphere. Spinning bubbles deform from spheres into flattened elongated ellipsoids. Spinning bubbles pulsate as small instabilities develop. Additional spinning produces a quivering dumbbell or peanut shape.

These shapes can also be obtained on earth in a flying water balloon just before the splashing disaster. A tossed spinning water balloon will first flatten along its spin axis. If the spinning speed is sufficient, the balloon will deform into two blobs connected by a narrow neck at the spin axis. In space a rapidly spinning bubble might divide into two bubbles of equal mass moving apart in opposite directions with equal amounts of momentum. This division occurs when surface tension can no longer hold the bubble together. Raindrops might break up within clouds in the same manner.

As a by-product of the fuel cells, the space shuttle's on-board water supply is rich in trapped hydrogen gas. As a water bubble spins, the heavier water molecules move to the outside and the lighter hydrogen molecules hover around the spin axis. A cylindrical hydrogen gas bubble forms inside the water bubble.

With a straw, Dr. Hoffman blew air into a water bubble. A 3-inch-wide water bubble grew to a 6-inch-wide water bubble with air trapped inside. The air-filled water bubble became much harder to control, more subject to air currents, and much more likely to collapse.

The bubble project aroused the curiosity of STS 51D payload specialist Charles Walker (who might also have felt the need to protect his Continuous Flow Electrophoresis System from the potential floating water bombs). The neutral buoyancy demonstrated by the floating bubbles actually causes the electrophoresis experiment to work. In zero gravity there is little settling and mixing of substances floating in liquids. Temperature and density differences do not result in currents that stir up a liquid. In space a very stable liquid medium can be provided for electrophoresis. Eventually this process might produce drugs of greater purity than any process on earth.

Once convinced that the electrophoresis equipment would survive, Charlie Walker joined Jeff Hoffman in the injection of a water bubble with strawberry drink. The pink liquid formed beautiful flow patterns inside the great swirling water drop. Soon the multicolored spinning bubble split. It seemed logical to put the strawberry bubbles back together. Blowing and nudging set the bubbles on a collision course. Instead of joining, the impact resulted in splashing strawberry bubbles flying everywhere. The insuing towel clean-up left Jeff Hoffman with strawberry-covered hands. In zero-gravity conditions, liquids form webs between fingers and clump in bubbles on hands. The red color made this strange sight even more bizarre. A day of soap and water washing brought the hands back to normal.

Skylab astronauts also completed interesting zero gravity bubble studies. They found that bubbles would spread out more readily on a metal surface and remain more spherical on plastic. They discovered that if a bubble were placed around a stretched string, the bubble would surround the string and elongate its shape at the points where the string crossed the bubble's surface. They also caused bubbles to collide. The bubbles would join if the air between them was pushed out of the way by the collision. If an air pocket remained, the bubbles would bounce apart. Once two bubble drops joined, they might still shatter apart if their combined speeds were too great before the collision.

Skylab astronauts also stretched bubbles apart and released them to make the droplets vibrate. Surface tension pulled the liquid drop back to a spherical shape. The liquid's momentum, however, caused the bubble to distort again in the opposite direction. Once again surface tension would pull the bubble back to a round shape and the oscillation would repeat. Vibrations lasting up to 30 minutes could be produced in a water bubble.

RESTROOM FACILITIES

Using the restroom also requires astronaut awareness of fluid properties. A private toilet area occupies a corner of the middeck. The toilet has the appearance of its terrestrial equivalent. A reading light and a hatch window for viewing the earth provide the comforts of home. A few features do make this toilet unique. The visiting astronaut must put his or her shoes in the foot restraints and buckle the seat belt to remain on the toilet. Without gravity, terrestrial water flowing motions will not occur. Flushing is accomplished by a fan that pulls wastes away from the user and sends them to the lower deck. A directed air flow blows liquid waste into a flexible hose that leads to a wastewater tank below. Wastewater is discharged at high speed into space, creating a stream of snow and ice particles that are eventually carried off by the solar wind. The STS 54 crew became the first to test a new microgravity toilet and rated it as an improvement on the one that the STS 51D astronauts used.

HAIR AND CLOTHING

Personal grooming habits must adapt to weightless conditions. Spaceflight hair styles for women offer new options. Hair can be worn short like the male astronaut hair styles. Longer hair can be restrained with a clasp or it can fly freely in all directions. Rhea Seddon had this choice with her long blonde hair. Susan Helms' short curly locks looked much the same in space as on earth. To check your hair style's spaceflight potential, hang your head upside down. Every lock that moves will stand straight out in space.

Susan Helms had no problems with her short curly hair, but made a discovery about shirts. She always chose loose-fitting shirts on earth, but found that loose-fitting clothes float around you in space, making you look and feel much larger. Well-fitting shirts work much better in microgravity.

NIGHT IN SPACE

Sleep is one of the few human terrestrial activities that is more easily accomplished in space. In weightless conditions, the air itself becomes the softest possible mattress. Four middeck support bunks provide privacy and keep floating sleepers stowed out of the way of working astronauts. Two bunks face upward in the traditional manner, one bunk is upside-down, and the fourth is vertical. In space, of course, these directions are meaningless. Sleeping bags with long front zippers surround each sleeper. Arms left outside are strapped to the bag at the wrist. Crew members on missions with a single work shift might ignore the bunks and choose instead to hook themselves to lockers in a corner of the middeck. Susan Helms wanted a confined place for sleeping and finally settled on one of the cockpit chairs where she could strap herself in.

The normal 24-hour sunrise-to-sunrise sequence occurs every 90 minutes as the shuttle completes one earth orbit. Crew schedules allocate 8 hours for sleeping during each 24-hour period whether it is dark or not.

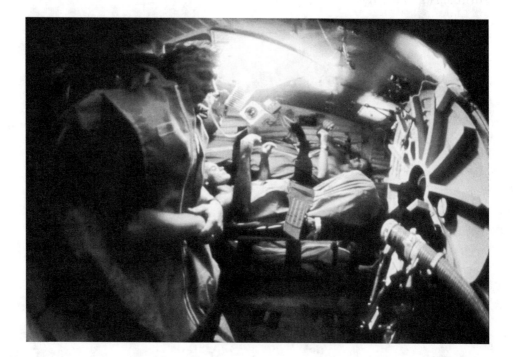

Eye shades and ear plugs are needed to give the illusion of night. Once eye masks are in place, sleep should soon follow. Getting accustomed to sleeping and floating can present problems for astronauts who tend to doze off. Without gravity, sleepy heads do not nod and then jerk to awaken their tired owners. Eyes simply close and sleep comes. A companion's gentle nudge or a question blared in a headset from Mission Control might be required to wake up the embarrassed astronaut.

CONCLUSION

Astronauts cherish their free moments in this weightless air-filled swimming pool called shuttle. They learn to do free-floating flips by swinging their arms and watching their bodies turn the other way to conserve angular momentum. Floating flips, weightless games of catch and fetch, times for frolic with spinning water balls, these are special experiences for each astronaut.

By mission's end, Jeff Hoffman mastered the art of making porpoise loops from flight deck to middeck and back. His enthusiastic acrobatics led Commander Bobko to reflect that their goal was to make 100 revolutions of earth, not 100 revolutions of the space shuttle. Each astronaut eagerly donated a few of these precious free moments to play with familiar toys and to share the experience of space flight with earthbound space enthusiasts of all ages.

4

Space acrobats

FOUR TOY ASTRONAUTS are acrobatic specialists: a hopping frog, a flipping mouse, a climbing astronaut, and a jumping grasshopper. Each behaves like a real astronaut might in the roomy environment of a future space station. Begin this space acrobatics adventure by constructing your own space grasshopper. All you need is a square sheet of paper to make a pet grasshopper. After experimenting with the grasshopper, you can predict how the other store-bought toy astronauts might perform in microgravity.

THE GRASSHOPPER

You will need

- One square piece of paper, approximately 3 inches on a side (a Post-it note works very well)

STEP 1 & 2   STEP 3

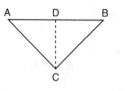

STEP 4   STEP 5

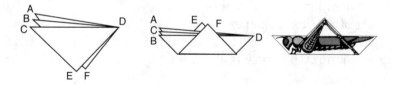

1. Fold the paper into a triangle. If you are using a Post-it note, fold the sticky side on the inside.
2. Make a crease along the middle of the triangle. Then open the grasshopper back up into a triangle. The crease runs through point C as shown on the drawing.
3. Fold up the corners labeled A and B until they touch C and the fold lies along the crease line. The shape is now a diamond or square.
4. Fold along the center crease from C to D. Fold so that the smooth side of the square is inside the fold and the flaps are on the outside. The shape is now a triangle again.
5. Corner D is the grasshopper's nose. Fold up the two tips at E and F to give the grasshopper his bent legs.
6. Place your grasshopper on a flat table and push its nose downward. Your grasshopper should flip. Make the creases very sharp and strike the very tip of its nose to get the best flip.

Do you think larger grasshoppers will hop higher or farther than your small grasshopper? Make a prediction and then make grasshoppers from paper squares of different sizes. Use the same kind of paper for each grasshopper. Experiment to find out which grasshopper jumps the highest and farthest. Make grasshoppers with thin wrapping paper and heavy card stock. Which paper stock works best?

The grasshopper game

With green construction paper, make a plant with long leaves as a target for your grasshoppers. See how far you can move away from your plant and still land your grasshoppers in it. You can also make lots of grasshoppers of different colors and use them in a board game. Each player gets points for every grasshopper that flips onto the plant. The farther away the grasshopper is, the more points that are scored.

Grasshopper physics

As you push downward with your hand, you cause the grasshopper to flip upward. The action is your downward push. The reaction is the upward push of the table on which your grasshopper sits. When you have made a grasshopper that flips well, try flipping it on different surfaces: table, floor, carpet, pillow, etc. Some surfaces give the grasshopper a greater upward push than others.

The size of a terrestrial paper grasshopper is very important. Your finger is supplying the action force that results in the reaction force that makes the grasshopper jump. This force is about the same whether you are trying to flip a small grasshopper or a large one. The large grasshopper has more weight which the jumping force must overcome. For this reason a smaller lighter grasshopper will usually jump higher and farther. Will size play a similar roll in the behavior of paper grasshoppers in space?

The grasshopper spins because of the downward push on its nose or tail to make it flip. This pushing also gives the grasshopper angular momentum. The grasshopper keeps spinning after it leaves the ground to conserve angular momentum.

RAT STUFF, THE FLIPPING MOUSE

Rat Stuff is a mechanical mouse that flips by leaning forward and then pushing backward with its feet. A gymnast on earth pushes off her toes in much the same way when she does a back flip. As Rat Stuff's feet push down, the surface pushes up, causing the mouse to leave the table. The mouse then flips around its center of mass. On earth the mouse returns to its feet.

Rat Stuff uses the same physics motions as the grasshopper does. The action of Rat Stuff's feet pushing on a table cause the table to push back in a way that produces the mouse's back flips. The grasshopper and mouse differ in their energy source. The grasshopper depends on a push from an astronaut. Wind-up toys such as Rat Stuff store energy in a spring motor that is wound up as the astronaut turns the knob. When the

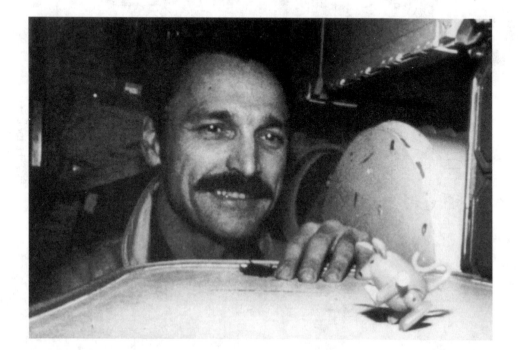

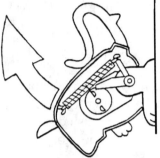

Cam

Flat spring motor

The cam mounted to the main shaft of the flat spring motor turns in a counterclockwise direction.

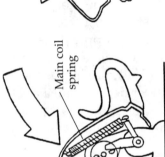

Leg armature

Main coil spring

The force of the cam's rotation against the leg armature causes the main coil spring to be stretched (storing energy) and the torso of the mouse to be pushed downward and forward.

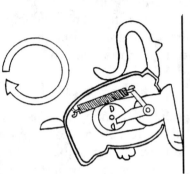

When the rotating cam reaches its point of disengagement with the leg armature, the energy stored in the main coil spring is released causing the torso of the mouse to be jettisoned backwards and upwards, momentarily overcoming the effects of gravity.

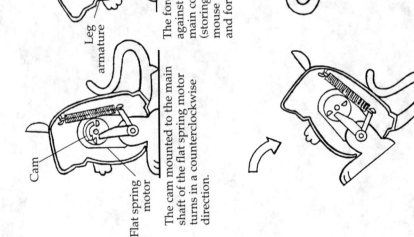

The backward and upward force that the main coil spring imparts to the torso causes the mouse to rotate around its own center of gravity.

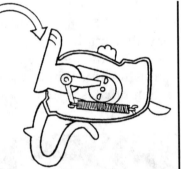

NOTE: The energy stored and released from the main coil spring has been adjusted to impart the proper amount of force to allow the mouse to land back on its feet.

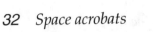

knob is released, the motor turns a shaft that makes the mouse tilt forward. The forward tilt stores energy in a stretched spring. At the proper moment, this spring is triggered. Its release pulls the feet and tail together with enough force to make the mouse flip.

To analyze mechanical systems, it is often necessary to perform toy surgery. As a basic rule of toy science, never operate on a toy that you expect to work again. Small toys have very tiny moving parts that tend to fly in all directions. Also pressed or melted together plastic parts might refuse to rejoin after the surgery. Broken toys definitely make the best patients. After determining how all moving parts move or refuse to move, you might actually be able to fix the toy. Occasionally bits and pieces from several similar toys can make one toy that works even better than its ancestors.

An autopsy of a deceased Rat Stuff reveals a flat spring motor containing a shaft with a cam attached. The cam's rotation against the leg joint causes the main spring to be stretched. The mouse stores elastic potential energy as it leans forward. When the cam turns far enough to release the leg, the energy in the spring pulls the body backward away from the feet. The mouse rotates around its center of mass. Its angular momentum is just enough to carry the mouse around and prepare it for a three-point landing using its feet and the bump on the bottom of its tail.

Two earth activities can help in predicting what the mouse will do. First, place the wound-up mouse on a carpeted surface and then on a pillow. Notice how the surface affects Rat Stuff's flipping motion. Next, wind up the mouse and watch its head move. When it is just about to flip, toss it into the air. What is the result of the flipping motion? This activity works better if one person tosses and the other observes. It might take several tosses to catch the mouse trying to flip in the air.

THE SPRING JUMPER

Walking and jumping depend on gravity to hold the toy or astronaut on a surface. The hopping frog experiences many of the same problems that Rat Stuff and the grasshopper face. When the frog's spring is compressed, energy is stored. This potential energy is turned into the frog's

kinetic energy when the suction cup releases the spring. The suction cup holds because there is air pressure on earth and in the shuttle's cabin that pushes in on the cup. This pressure is greater than the partial vacuum inside the suction cup. Newton's third law predicts what will happen when the spring releases. The frog's base pushes against the table and the table pushes the frog upward. Once again it is action/reaction at work.

Imagine pushing the frog's spring together and setting it on a hard, flat, level table. Now imagine placing the frog on a soft, flat, level carpeted floor, and then on a very soft, flat, level pillow. If you have a spring jumping toy, this makes a very good experiment. It is important to realize the effect of the surface on the strength of the jumping force.

In many spring jumpers, the stand can be separated from the spring. If this can be done, then the spring jumper can be launched by sticking the suction cup on a table. With the mass of the stand removed from the spring jumper, the force of the spring can produce a much higher hop.

THE CLIMBING ASTRONAUT

The climbing astronaut is an adaptation of an old-fashioned toy. Strings run through the astronaut's hands and connect to a rocking bar behind the

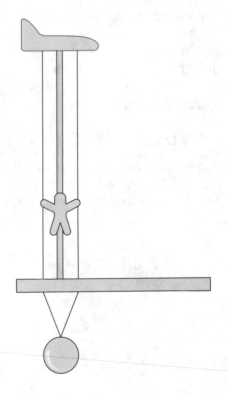

shuttle. Below the handle, these strings connect to a swinging ball. As you swing the handle, you cause the ball to swing around. As the ball swings, it pulls on the strings one by one, causing the wooden astronaut to climb upward to the shuttle. Study the drawing of this toy. It is easy to construct if you have equipment to work with wood.

You can experiment to see if gravity has much effect on the toy climber. If you swing the handle vertically, the climber climbs. If you swing the handle horizontally, the climber climbs. If you take the climber off and put it on the strings backward, it will climb down the strings as you swing the handle. It's centripetal force that makes the toy astronaut climb. The ball tries to move in a straight line (inertia principle), but the strings cause it to turn in a circle. The strings provide the centripetal force. This force is greater on the string that is farther from the ball. The extra tension on this string pulls the string through one hand of the astronaut. When the ball moves to the other side of its path, the other string exerts more force and the other hand climbs. As you imagine this toy working on earth, try to think of problems it might experience in microgravity.

SPACE RESULTS: GRASSHOPPER

Each acrobat has a different story to tell. The grasshopper is one of the toys that the astronauts did not have time to demonstrate. Instead, the grasshopper's experiments were redesigned to be done in the 20 seconds of freefall that can be produced during each dive of the KC-135 airplane. As everyone in the cabin began to float, Linda Billica placed the grasshopper in the air. It floated motionless until she pushed down on its nose. Then the grasshopper began flipping and moving downward. It continued the spinning motion around its center of mass until the period of freefall ended. Then the grasshopper and other passengers dropped to the floor of the airplane. Although the grasshopper did not push off of a surface, it did perform its moving and spinning motion across the airplane. Until the plane stopped its dive, there was no force to pull the grasshopper toward a surface.

SPACE RESULTS: RAT STUFF

Rat Stuff's performance was much more dramatic than a simple jump. A flip is far more graceful in space, but a bit more difficult to attain. Like the spring jumper, Rat Stuff does not stick to a flat surface in space long enough to push off. Don Williams and Susan Helms tried different strategies to help Rat Stuff flip. Don Williams administered a little hand cream to each foot area, but the mouse's feet still slipped. He increased the amount of hand cream to give the mouse stickier feet. Finally a layer as thick as a pencil eraser held the mouse in place until he jumped.

With surface tension from the hand cream holding his feet in place, the wound-up mouse waited obediently for his chance to perform. With a powerful kick, he sprang off the wall. The initial flip gave this tiniest astronaut acrobat enough spin to tumble over and over as he sailed across the cabin. His was the fate of all novice astronauts. Accustomed to gravity's downward pull, his pushing feet exerted too great a force for the graceful, slow motion world of microgravity. Without tiny mouse toeholds for stability, Rat Stuff floated about like his full-size companions. On earth, each flip ends as Rat Stuff makes a three-point landing using his feet and tail. In space no force stopped the flipping mouse until his trainer made a wild grab for the sailing miniature astronaut. In true space fashion the flying Rat Stuff also received a tiny strip of Velcro on his feet courtesy of Charlie Walker. Only four Velcro hooks were needed to hold the mouse to the Velcro pile and still allow him to flip.

In Rat Stuff's second mission, Susan Helms developed an ingenious deployment strategy for the flipping mouse. She released Rat Stuff in her hand. Three of his flips went flawlessly as he left her hand in a flipping motion. The pushing action of Rat Stuff's feet was matched by a reaction push from Susan Helms' hand. Her hand was less rigid than the locker used in the first mission and the resulting series of back flips was slower and much more graceful and elegant. On the ill-fated fourth attempt, Susan Helms' hand slipped from beneath Rat Stuff's feet and the mouse kicked uselessly in the air until she rescued him. In response to an idea from students, Susan Helms also taped the mouse's feet to a crew flight book. Then the mouse kicked in vain. His small feet could not begin to move the massive book.

SPACE RESULTS: SPRING JUMPER

Commander John Casper operated the hopping frog. Carefully he squeezed the spring and held it closed with the suction cup. As expected, the suction cup worked well in microgravity because the cabin air is still at earth surface pressure. Then he placed the frog carefully in his open palm, trying not to give the creature the gentle nudge that would send it floating into space before it could hop. When the suction cup released, the spring pushed the frog's base against John Casper's hand and the frog sprang forward. Within a moment it sailed straight across the cabin and into the storage locker doors. It bounced off the lockers and downward toward the astronaut's feet. Next John Casper released the frog backward with its head against his palm. The result was much the same although the flying frog's speed was slower. It seemed that its head was held a bit more firmly in the astronaut's hand and the creature was not quite as aerodynamic in the upside-down position.

SPACE RESULTS: CLIMBING ASTRONAUT

The climbing astronaut rode into space but was not tested. Instead, Linda Billica put the climber through its paces on the KC-135. Her problems came with the ball, not the climber. It was difficult to keep the ball moving in circles without tangling up the strings in the handle. By keeping the ball below the handle, gravity definitely helped the operation of this toy. Whenever she could control the ball, the astronaut would climb along the strings.

CONCLUSION

Both Rat Stuff and the spring jumper push on a surface to jump. The astronauts must produce this surface contact for either toy to work. The grasshopper needs a tap on its nose to make it flip on earth or in space. In space, this tap makes the grasshopper move and spin as it does on earth. Of these three toys, the grasshopper's behavior changes least in microgravity.

The flipping mouse, hopping frog, and jumping grasshopper show how human motions must be adapted for space flight. An acrobat would have to relearn her routines in space. But the trade-off would be well worth the effort. Instead of doing just one flip off the table, the space Rat Stuff was able to flip until he hit another wall. Any gymnast in a large space cube could perform feat after feat while bouncing from wall to wall. Slow graceful space motions are natural, even for nonacrobatic astronauts. Fast-moving jumps and twists would be challenging, especially the recovery, without gravity's stabilizing pull.

The antics of these space acrobats are more than simple physics demonstrations. They give us a scale for human behaviors in a large space station of the future. Shuttle astronauts run into walls when they attempt tricks in space, but the shuttle provides plenty of room for these miniature toy astronauts.

5

Space movers

TO GET ACCUSTOMED to a floating environment, astronauts train in a large pool of water at the Johnson Space Center. Inside their space suits, they can simulate what it is like to walk in space. Three of the toys in the STS 54 mission were designed to move in water: the fish, the frog, and the submarine. How well will these earthly swimmers move when they are floating in air instead of water? Will they move as well as the three toys that are designed to move in air: the flapping bird, the double propeller, and the balloon helicopter.

All toys that move through air or water use the action/reaction principle. In some way each toy pushes backward against the air or water. In return, the water or air pushes the toy forward.

THE SWIMMING FISH

The space fish uses its tail to propel it through water. The tail flips from side to side as the fish swims forward. At each swish, the tail pushes water backward and receives a forward push from the water. As you wind the knob on the fish, a spring is wound up inside. As the spring unwinds, it turns four gears. The last gear turns a wheel with an off-center knob, and the knob pushes the fish's tail back and forth. Will the tail be as effective as it moves air? How could the astronauts make the fish's tail work better?

THE SWIMMING FROG

The space frog uses its feet-like flippers. It pushes backward against the water with both feet simultaneously. As the feet push backward, the frog glides forward. The spring device inside the frog is just like the one in the fish. In the frog, the knob pulls the frog's legs together and then pushes them apart. The frog's legs move back and forth at about the same rate as

the fish's tail flaps from side to side. Will the frog's two smaller feet be as effective as the fish's larger tail in pushing it through the air?

THE SUBMARINE

The space submarine uses a rapidly turning propeller to push the water backward and move the sub forward. Although the propeller blades are much smaller than the frog's feet or the fish's tail, they turn at a much faster rate. The sub has the same wind-up mechanism as the other two swimmers, but in the submarine the fourth gear turns the propeller shaft directly. The submarine also has stabilizer fins on either side of its body. These keep the submarine from twisting when the propeller turns. Do you think these stabilizers will be as effective in air as they are in water?

THE DOUBLE PROPELLER

This is a real space toy that will not work on earth. The toy is made from two balsa wood propeller airplanes with plastic rubber band-powered pro-

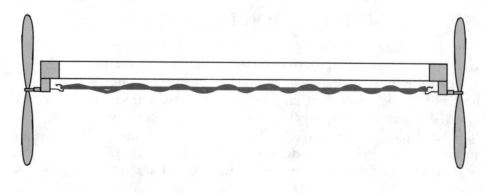

pellers. A propeller is placed on each end of one of the planes. The wings and tail rudders are removed. A rubber band connects the two propellers. When the rubber band is wound up and released, it turns both propellers at the same time. Each propeller is designed to push the air backward causing the plane to go forward. However, the two propellers mounted on either end of the fuselage push the plane in opposite directions as they turn.

As the rubber band is wound, energy is stored in the system, which is released as the propellers spin. Without wings the double propeller cannot support itself in air, so you cannot watch it move on earth. The astronauts also carried a third propeller with blades that were twice as large as the standard propeller. When the larger propeller replaces the standard propeller, it might make a difference in how the double propeller flies. We know that the larger propeller pushes more air, but also offers more resistance to spinning. Since both propellers are attached to the same rubber band, the smaller one will probably spin more quickly. Perhaps the astronauts can discover if larger blades or a faster turning rate work better.

THE FLAPPING BIRD

The flapping bird is designed to fly in air and uses its flapping wings to overcome the downward pull of gravity as it moves forward. The bird has two wings that move up and down as it flies. The power for the wings is in a rubber band that is wound up about fifty turns before the bird is released. As the rubber band unwinds, it turns a cam that pulls the wings up and down. The bird is much lighter than the space movers described previously. For this reason, it can fly on earth. In microgravity, the difference in mass is much less important. On earth, the bird's wings push the air downward so the air pushes the bird upward and gives it lift (the action/reaction principle again). Without this upward push from the air, the bird would glide to the ground. What effect will these flapping wings have in microgravity?

THE BALLOON HELICOPTER

The sixth space mover is a balloon-powered helicopter. In this toy, air rushing from an inflated balloon causes the helicopter to fly. The air forced from the balloon moves through the wings and leaves at an angle at the tip of each wing. The air nozzle on each wing tip is angled downward and backward. In this way the air flowing downward causes the helicopter to climb upward. This backward motion causes the blades to turn in a counterclockwise direction when viewed from above. As the blades turn, the helicopter climbs. The helicopter rises until the balloon is

deflated. The momentum of the escaping air moving downward is balanced by the momentum of the balloon moving upward. Momentum is conserved as the air escaping in one direction causes the helicopter to move in the opposite direction.

Before predicting which space mover will do best in air, you can experiment with different swimming strokes that you use in water. You might also play with releasing a balloon and watching how it moves in air.

All six movers push backward on the air in order to move forward (Newton's third law). But each uses a different mechanism for propulsion. Before you read on, predict how the movers will work and rank the six movers from fastest to slowest in their freefall flights.

SPACE RESULTS: SWIMMING FISH

Both John Casper and Susan Helms played with the fish. The astronauts carefully released the fish and tried not to provide the push that would make it tumble. Once the deployment technique was perfected, the fish's tail moved the fish slowly across the cabin. An enlarged tail fin worked wonders for the floundering fish. With paper added, the new tail was over twice as large as the old tail, and the improved fish wiggled effectively across the middeck.

SPACE RESULTS: SWIMMING FROG

To the surprise of kids and physicists, the frog was the slowest space swimmer. Carefully Susan Helms released it from her hands. Each time she opened her hands, she accidentally gave it a slight push and caused it to drift and tumble across the cabin. Although the frog's legs moved back and forth, they were unable to push enough air backward to move the frog forward. Off-camera the astronauts gave the frog bigger flippers, and it did finally swim forward very slowly.

The unfortunate frog shows an important difference between air and water. Think about waving your hands in water and in air. You must push much harder, exerting much more force, to move water. Your action force is much greater as you move your hand through water. The resulting reaction force from the water is also much greater in pushing you forward. The frog could easily push enough water with its feet to move forward, but it could not move enough air with its feet to produce an effective reaction push from the air.

SPACE RESULTS: SUBMARINE

The propeller-driven submarine was a better swimmer in air than the fish, frog, or bird. But the submarine's technique had a surprise side effect. As the propeller turned in a clockwise direction, the submarine began spinning in a counterclockwise direction. The turning propeller has angular momentum in one direction. To conserve angular momentum, the freely floating submarine begins spinning in the other direction. In water, the submarine's stabilizer fins keep it from spinning. In air, these fins prove useless and the submarine turns faster and faster. Even with this spinning, the submarine's propeller did move it across the cabin.

At the suggestion of a student, Susan Helms taped a writing pen to the propeller. The pen began spinning quickly when the submarine was released, and the submarine continued to spin in the opposite direction. The submarine did move forward, even with the spinning pen attached. John Casper added little paper extensions to each propeller blade and made each blade about twice as big. With this design, the submarine's forward speed more than doubled.

SPACE RESULTS: DOUBLE PROPELLER

A real helicopter has a second propeller on its tail. This propeller stabilizes the helicopter and keeps it from turning. The propellers turning in opposite directions do the same thing for the double propeller toy. While floating in air, this toy does not twist like the submarine. The STS 54 astronauts ran out of time before they could demonstrate the double propeller. The toy was flown on KC-135 flights before and after the shuttle flight to get a good idea of what the toy would do. Although its performance lasted only 20 seconds at a time, the double propeller turned out to be a very elegant toy. The propellers pushed the air in opposite directions, keeping the toy in the same general location in the airplane. Both blades push against the air in the same angular direction. As a result, the double propeller toy began to spin around its center of mass. This toy was demonstrated only twice. To determine what the toy is really doing, it must be observed in space for a longer period of time.

SPACE RESULTS: FLAPPING BIRD

For all of its wing span, the bird had as much trouble moving as the fish did. Most observers predicted that the bird would be very successful in

space. After all, the bird was designed to move in air. But without any downward pull, the flapping wings sent the bird into back flip after back flip across the cabin. Flying and flipping into a corner finally stopped the bird's wild flight. A real bird in freefall spreads its wings and soars. This mechanical bird could not shut off its wings and glide. After two unsuccessful flapping flights, John Casper pushed the bird gently across the cabin without winding the rubber band. In level flight, the bird made a wonderful glider.

SPACE RESULTS: BALLOON HELICOPTER

If the submarine's small blades could send it flying across the cabin, the balloon helicopter was destined to be a great success. The air escaping from the inflated balloon caused each wing tip to turn and rise. The escaping air from the wings carried the helicopter upward at over twice its rising speed on earth. When the helicopter reached the far wall of the cabin, the crash knocked off the balloon. Rocket power is definitely the best mover in microgravity, especially when the object has little mass.

Momentum conservation makes rockets work. As air moves backward out of a rocket, the rest of the rocket must move forward to conserve momentum. The faster the air is moving, the faster the rocket moves in the opposite direction. The inflated balloon pushes the air quickly through the blades and creates a superspeed helicopter.

To maneuver the shuttle, astronauts fire small maneuvering rockets located in the nose and tail of the shuttle. When these rocket engines are fired facing one direction, the shuttle turns in the other direction. Astronauts wear smaller rockets in a free-floating chair called the Manned Maneuvering Unit (MMU). Small rockets can move a single astronaut around the shuttle cargo bay without a tether. The astronaut uses these rockets to return to the air lock. Neither STS 54 or STS 51D carried the MMU, although both missions conducted activities in the cargo bay.

6

Space flyers

THE PAPER AIRPLANE is the simplest of all flyers. It uses air to carry it on a glide path toward a gentle landing. Gravity pulls each paper glider downward, but it is hard to tell how much effect gravity actually has on the flyer's path. For the STS 54 flight five different paper flyers were designed to do different things in the air and perhaps different things in microgravity. Two are planes: the spiral plane is a traditional craft like the one that Senator Garn flew in the STS 51D mission. The ring-wing glider is a very strange looking craft that works on earth and may have very unusual behaviors in microgravity. The remaining three aircraft produce a spinning motion as they sail through the air. The one-blade maple seed and two-blade whirly bird both spin around a center axis. The four-blade boomerang changes its spinning orientation as it flies. You can assemble each of these flying craft, try it on earth, and predict its behavior in freefall.

THE BERNOULLI BLOWER

The STS 54 astronauts also selected one toy that illustrated how the flow of air around an object could affect how the object moves. This toy is called a Bernoulli blower. The store-bought version has a pipe with a basket on the end. A ball rests in the basket. When you blow on the pipe, the ball hangs in the airstream above the pipe. You can make a very functional homemade blower from a ping-pong ball and a straw with a bend.

Place the long end of the straw in your mouth and bend the other end until it is vertical. Hold the ping-pong ball above the bend and blow. The air escaping from the straw will float the ping-pong ball. You can move your head slowly and the ping-pong ball will stay above the straw. You might even be able to tilt your head and keep the ball in the air stream. To save your breath, you can use a hand-held hair dryer to support the ball.

The Bernoulli blower illustrates the Bernoulli principle. The air pressure is lower in a faster-moving air stream. When you blow below the ball, you create a stream of air flowing on either side of the ball. The ball stays in this air stream because the air pressure beyond the stream is greater. If the ball drifts to the side of the air stream, the slower moving air beyond the air stream pushes the ball back. Watch as the ball floats above the blower. How must you adapt this toy for microgravity?

THE RING-WING GLIDER

Follow these steps to construct a ring-wing glider.

1. Fold a sheet of paper along a line connecting opposite corners. You might want to draw this line on the paper to make folding easier.
2. Make a roll about 1 inch wide along your fold line. Then roll over the paper for a second time.
3. Connect the two ends of your roll to make a tube. You can push the ends into each other and they will usually hold. Use a small piece of tape if the ends will not stick together.
4. Hold the tube with the "V" between your fingers. Then flick the tube forward. Watch what happens.

With some practice, you can make your tube soar and loop. It doesn't have to fly in a straight line. Your goal is for the tube to make a complete loop and land near your feet. Notice how the tube climbs, stalls, and then falls.

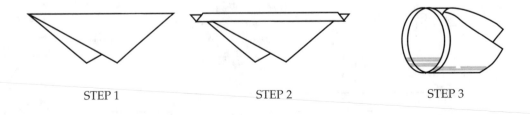

STEP 1   STEP 2   STEP 3

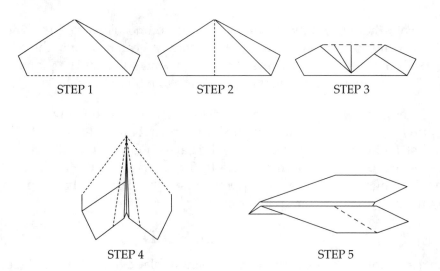

STEP 1   STEP 2   STEP 3

STEP 4   STEP 5

THE SPIRAL PLANE

Follow these steps to construct a spiral plane like the one carried into space.

1. Fold a sheet of paper by connecting opposite corners.
2. Now crease the paper by folding it together at the tip where the corners meet.
3. Fold over the tip along the crease.
4. Fold over the two ends along the crease line to make wings for the plane.

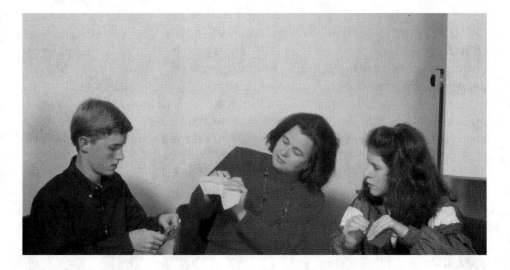

5. Fold the wings upward along the crease line. Then fold the wings downward leaving a central area where you can hold onto the plane.
6. Throw the plane forward. Adjust the wings so that the plane will spiral. Then flatten the wings so the plane flies straight.

Experiment with your plane before predicting how it will work in microgravity. First fly the plane forward, then turn it around and try to fly it tail first. Notice what happens. Make the flaps about 0.5 inch (about 1 centimeter) deep and 1 to 2 inches (2.5 to 5 cm) wide on the backs of the plane's wings. Observe what happens when the flaps are bent up, bent down, and with one bent up and the other bent down.

THE MAPLE SEED

The real maple seed is a very ingenious single-blade flyer. The maple seed pod has one large blade. When the seed falls from a tree, it spins all the way to the ground with the seed pod below the blade. Gravity pulls the maple seed downward, but what are the forces causing the seed to spin? You can make a maple seed to find out.

A paper maple seed was flown on the shuttle. It was folded using a complicated origami pattern. The same effect can be achieved by cutting these patterns from blank paper. You can experiment with different weights of paper to see which works best.

After cutting out the pattern, attach a paper clip to the paper as shown. Then curve or twist the blade slightly so it will catch the air. Drop your maple seed and watch for a spinning motion. If it does not spin, adjust the position of the paper clip or the curve of the blade. Once it spins, experiment with throwing the maple seed at different speeds. Experiment with different blade angles to create maple seeds that fall slowly and ones that fall quickly. Predict what will happen when the maple seed is thrown in microgravity.

Notice what happens to the seed as it falls to the ground. Its streamlined forward flight helps the seed ride the wind far from its parent tree. Once the wind slows down, the seed begins to drop. As it falls, its blade catches the air causing the maple seed to turn. This turning motion stabilizes the maple seed so that it lands with the seed pod down. It also slows the fall for a gentle landing.

THE WHIRLYBIRD

You will need a square sheet of paper about 6 inches on each side for each whirlybird. Use the illustrations as you follow the folding steps.

Follow these instructions to make a fast twirler.

1. Make creases in the paper connecting opposite corners. Then open up your paper.
2. Fold two opposite corners inward to the center of the paper.
3. Fold inward again toward the center over the folds you have just made. The result is a long thin piece of paper.
4. Fold inward again toward the center line to make an even thinner piece of paper.
5. Fold this paper in half with the folds on the outside.
6. Fold out the wings on both sides of your whirlybird.
7. Use a small piece of tape to hold your whirlybird together if needed.
8. Release your whirlybird and watch it spin toward the ground.

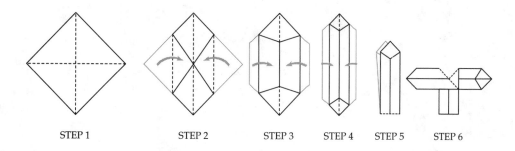

STEP 1   STEP 2   STEP 3   STEP 4   STEP 5   STEP 6

Follow these instructions to make a slow twirler.

1. Repeat steps 1, 2, and 3 from the fast twirler instructions.
2. Fold the paper in half with the folds on the outside.
3. Fold out the wings on both sides of your whirlybird.
4. Use a small piece of tape to hold your whirlybird together if needed.

Stage a twirler race once both twirlers are made. Predict which twirler will fall to the ground first. Drop both twirlers from the same height at the same time. Notice which one reaches the ground first. Notice which one is a better spinner. Repeat several times to be certain that your result always happens.

THE BOOMERANG

When you send a boomerang spinning through the air, it is supposed to come back to you. Traditional two-blade boomerangs need a space about as big as a football field to turn around. The astronauts have only the shuttle's middeck, which is about 12 feet (about 4 meters) wide. For the astronauts, a small four-blade boomerang was designed out of card stock. Follow these instructions to make a boomerang like the astronauts used.

1. Use the pattern on page 48 to cut the exact shape out of a piece of card stock.
2. Notice that the card stock curls.
3. Hold your boomerang vertically by one blade and throw it forward with a spinning motion. The slight curl in the card stock should be to the left if you are right-handed and to the right if you are left-handed.
4. With a little practice, your boomerang should return every time.

As you toss the boomerang, decide if the air is making the boomerang turn around or if gravity is playing a part. Does the boomerang fall as it returns? Does the boomerang's spin change during flight. Is the spin changing in direction, angle, or rate of spin? Is this change caused by the air or by gravity?

SPACE RESULTS: BERNOULLI BLOWER

The Bernoulli blower received its flight experience before the STS 54 mission lifted off. It was tested along with many of the other toys in a KC-135 airplane flight. The flight version of this toy contained a piece of thread linking the ball to the blower. The KC-135 flight verified that without this thread the ball would just fly away. Gravity is critical in keeping the ball close to the pipe. Without gravity or a thread, the ball will rise in the air stream and float away. The thread keeps the ball in close, but it was difficult to observe in the few seconds of freefall available in each dive of the KC-135.

You can experiment on earth with the thread modification. Glue the thread to your ping-pong ball and tie the other end around your straw. When you blow, notice that the thread keeps the ball from turning and increases the ball's tendency to move out of the air stream.

SPACE RESULTS: RING-WING GLIDER

The ring-wing glider accompanied Senator Garn's standard paper airplane into space and went back into space on the STS 54 flight. The only video was shot on a test flight on the KC-135. On earth, this glider often climbs, stalls, and drops into a loop. Gravity causes the glider to drop. The KC-135 crew had a difficult time launching the ring-wing glider. The flicking technique that causes the plane to glide through the air on earth results in a plane tumbling quickly to the floor of the KC-135. Without a downward pull, the plane would not demonstrate the glide-stall-loop behavior that it does on earth.

SPACE RESULTS: SPIRAL PLANE

The performance of all the space gliders exceeded earth-based standards. A gentle push sent Senator Garn's airplane gliding through the middeck. Without gravity's downward pull, the plane soared straight across the cabin. It could be thrown at any speed and in any direction for an elegant flight.

Plane design is important, even in space gliding. When Senator Garn tossed his plane tail first, it tumbled about, showing no gliding motion. In response to student questions, Senator Garn also discovered that a paper airplane will bounce off the shuttle's walls and can be propelled just by blowing on its wings. When the plane was released with no push at all, it twisted and gently drifted in response to circulating air currents in the shuttle's middeck area.

Greg Harbaugh's plane had to be refolded from its flat stowed shape. When doing this, the wings were positioned flat instead of tilted upward. As a result, the plane did not spiral as it does on earth, but flew just as Senator Garn's plane had.

Senator Garn did not have the first space paper glider. Astronaut Dick Scobee flew a dart-shaped paper airplane in the spring of 1984. His plane also flew effortlessly with an elegant banked curve caused by the tilt of the wings. This airplane flight came at the request of a group of Houston-area students. Astronaut Scobee was the commander for *Challenger's* last flight.

The performance of both planes shows that objects must be aerodynamic to fly through the air—even in space. Without a downward pull, we can see more clearly how the shapes of these planes affect their motions. Bernoulli's principle gives an airplane its lift. If the distance traveled by the air over the top of a wing is longer than the air's path under the wing, the plane will rise. Flaps allow us to adjust the length of this air path, causing the plane to rise, fall, or make a turn.

SPACE RESULTS: WHIRLYBIRD AND MAPLE SEED

The spinning flyers were very successful in microgravity. On earth, gravity pulls the maple seed and whirlybird through the air as they fall. In space astronauts have to throw them through the air to produce this spinning motion. The whirlybird's blades push against the air at an angle as the whirlybird moves through the air. The faster the bird moves, the more air the blades push. As the blades push in one direction, the air pushes back in the other direction. The result is the spinning motion of the whirlybird.

The property that makes a maple seed spin is called *autorotation*. If the speed is correct, the blade catches the air and causes the maple seed to spin as it moves through the air. The maple seed's center of mass is at one end of the blade, while the center of lift for the blade is near the center. The blade's shape causes it to catch the air and spin.

The large blade pushes the air in one direction, causing the seed to spin around in the other. John Casper discovered that the speed of a maple seed is critical. About 5 minutes were required to master the art of maple seed throwing. First, John Casper threw the maple seed too fast and it sailed like a pencil across the middeck. Then he began to grab the maple seed at the base of the blade and throw it with the pod forward. The maple seed began to spin. The real spiraling effect did not occur until he threw the seed very slowly at about the speed of a real maple seed as it flutters to the ground. In space the maple seed cannot be dropped like a tree drops a real maple seed on earth. But once it receives a gentle push, the maple seed becomes a stable and beautiful flyer.

SPACE RESULTS: BOOMERANG

The boomerang's success on earth and in space comes from aerodynamics and angular momentum. On earth the boomerang does not return unless it is spinning. The faster it spins, the more quickly it returns. Its blades act like propellers pushing back against the air and propelling the boomerang forward. Because of the spin and the curvature of the boomerang, the orientation of the blades is always changing.

On earth, if the paper boomerang is thrown vertically with the blades curved inward, the boomerang will always return in a horizontal orientation. In space, this same throwing action will also produce a boomerang that curves back toward the thrower, but the effect is not nearly as strong. Although John Casper had perfected the boomerang throwing technique on earth with his own children, it required several throws to make the boomerang perform well in space. On earth, gravity pulls down on the boomerang, causing it to change orientation as it falls. The inward-curving blades have the same effect, but the turning does not happen as quickly without the downward pull. During one boomerang flight, the boomerang flew a curving path across the middeck and into the airlock. This interesting feat was never repeated on camera.

The faster the blades turn, the more quickly the boomerang returns. This effect happens on earth and in space. In space the boomerang travels in a straight line if thrown very slowly. There is no throwing technique that will cause the boomerang with curved blades to travel in a straight line on earth.

If the boomerang blades are curved outward away from the thrower, the boomerang will turn to horizontal flight, but will fly away from the thrower. Commander Casper could also cause this behavior by changing the curvature of the blades. He also tried to throw the boomerang beginning with a horizontal orientation. When thrown in this way, the boomerang also arcs away from the astronaut and out of the camera field.

7

Space wheels

THE UBIQUITOUS AUTOMOBILE rules the earth's surface, but meets its match in space. The problem is simple: in space its wheels will not roll. You can spin them, push them, and wind them up, but to no avail. The wheels just spin in place, providing no contribution to the car's forward motion.

Wheels were invented as friction fighters. Earthlings rely on wheels to move big, bulky things across the planet's surface. The friction force increases with the amount of surface area. Just ask any child slipping down a playground slide—the less body that touches the slide, the faster the ride. Wheels reduce surface area contact drastically. Sliding friction is also much harder to overcome than the rolling friction of an object on wheels.

Friction develops as gravity pulls down on the object to be moved. As the wheels turn, they generate a backward action force. The forward reaction force from the ground pushes the car forward. In microgravity, there is no downward force to hold the toy car's wheels on the table. Without forceful contact, there is no friction and turning wheels spin in place. Terrestrial car owners sliding on a wet or icy street quickly recognize the uselessness of spinning wheels.

THE FRICTION-ENGINE CAR

The toy car is so familiar to people of all ages that it makes an ideal toy to use in space. If you have a toy car with a pull-back friction engine, spend some time studying how it works. Then make predictions about what the car will do in microgravity.

Toy cars with tiny motors often have ingenious gear designs. In many of these cars, the frame can be lifted off for a better view of the motor assembly. Two gears encircle the rear axle. When the car rolls forward, the motor turns these gears in the same direction. To wind up the motor, you push down on the car and roll it backward or forward. This action engages two more gears that make the axle gears turn in opposite directions, causing one gear to slip on the axle. These gears now touch the motor and wind its two gears in opposite directions. Releasing the car disengages all gears except the one from the motor, which turns a gear on the axle, causing the wheels to turn so that the car moves forward. As the car moves forward, the mechanical potential energy you stored in the engine is converted into the kinetic energy of the moving car.

If you have a small toy car with a friction engine, you can experiment with how it works best. Roll the wheels back and forth while pressing down on the car. When you hear a clicking sound, the engine is fully wound. Release the car on different surfaces (wood, concrete, carpet, etc.) and discover which is best for the friction engine. Now predict how a toy car will work in space where there is no friction to hold the car to a surface.

Finally, place your wound-up car in a loop. You can use a paint bucket or large bowl with vertical sides as your loop. Or you can purchase loop track to use. Hold the loop or bucket very steady. Release the car inside the loop. Count how many circles it makes. As the car circles in the track, it tries to travel in a straight line—as described by the inertia principle. The track pushes the car inward, causing it to move in a circle. This inward push is centripetal force.

If possible redesign your track to make a space slightly wider than the car is long. Close up the loop and release the wound-up car. Once the car has completed one turn, open the loop and see what happens to the car. Does it jump the gap and continue moving in circles?

THE COME-BACK CAN

The come-back can is a homemade toy with unusual behaviors on earth and in space. As a rolling toy, it faces many of the same challenges experienced by the friction car. A come-back can is easy to make. You can find all of the construction materials you need around the house, except for the fishing weight which you can get at a local hardware or sporting goods store.

To make your own come-back can, you will need

- One can that rolls and has ends (aluminum soft drink can, coffee can, 2-liter drink bottle, etc.)
- ½- to 1-oz. fishing weight
- A selection of rubber bands
- Large and small paper clips
- A nail or sharp knife
- Adult help for all kids making the cans

The exact procedures you use will depend on the can you have chosen. These instructions describe how to make the come-back can from an aluminum drink can. Modify these instructions to fit the can you have chosen. The come-back can used by the astronauts is clear plastic and shows the motion of the weight and rubber band inside.

1. Empty the can. An adult must punch a hole in the bottom with a nail or knife.
2. Attach the lead fishing weight to the middle of the rubber band. Tie it in such a way that the weight does not spin around the rubber band.
3. Open up a large paper clip to use as a hook.
4. Wrap one end of the rubber band around the pull tab of the drink can.
5. With the large paper-clip hook, pull the other end of the rubber band through the nail hole in the bottom of the can. Use a small paper clip to hold the rubber band in place outside the end of the can.
6. Wind up your come-back can by rolling it on a smooth surface in one direction. After it is wound up enough, it will begin coming back when you push it forward.

Watch as your come-back can rolls forward, comes to a stop, and then rolls backward. Notice how its speed changes as it moves forward. Compare the speed of the can at the time you push it and when it stops moving forward. What is causing the speed to change? Why does your come-back can move backward? Experiment with your come-back can. What can you do before you push it forward to make it come back farther? Release your can without a push on a long downhill slope. Mark where the can stops, and record how far uphill it comes back. Do come-back cans come back farther on level ground or moving uphill?

The come-back can is a toy that uses gravity to hold it to the earth's surface. When you roll the can forward, the weight hanging down winds up

the rubber band and the can slows down. The kinetic energy of the turning can is turned into the potential energy of the wound-up rubber band. When the can stops, the rubber band starts to unwind and the potential energy is converted back into kinetic energy as the can rolls backward.

SPACE RESULTS

In space there is no force to hold the car or the come-back can against the table and no friction to allow the action/reaction to happen. Instead of moving along the locker, the car's wheels spin uselessly as the car floats in the air. Both Jeff Hoffman and Mario Runco discovered that a toy car has no use for wheels. Don McMonagle came to a similar conclusion with the come-back can.

SPACE RESULTS: FRICTION-ENGINE CAR

With a circular track, the friction car had an interesting story to tell. Both Jeff Hoffman and Mario Runco used the circular track. A circular track 1 foot (30 centimeters) in diameter gives the space car a chance to perform. In accordance with the inertia principle, the car tries to move in a straight line and pushes against the curved track. The track pushes in on the car to provide the centripetal force that makes the car turn.

When the motor is wound and the car is released along its circular track, its wheels push against the track and the car rolls forward. This circular motion happens in space and on earth. On earth and in space, the track can be tilted in any direction and the fast-moving car will continue to roll. Only the running down of its motor causes the car to stop. On earth the car falls from the track to the ground. In space the car comes to a gradual stop. A gentle nudge on the track sends the car drifting out into the cabin.

On earth it is also easy to rock the car back and forth along the bottom of the track until it has enough speed to circle. This track movement will send the car racing in circles. By continuing to rock the track, you can keep the car rolling around—even when the track is tilted in any direction. In space there is no gravity pull to help start the car rolling. With a little practice, however, both astronauts could give the car a quick sling start along the track.

In the STS 54 mission Mario Runco had much more time to spend playing with the car and track. He experimented with holding the track very steady and with letting the track sway from the motion of the circling car. He quickly discovered that the car circled longer in the track when the track was held rigid. The swaying track was taking energy from

the car. He discovered that he could get the car spinning in the track no matter where he released the car inside the track. He also found out that it did not matter at what angle he held the track relative to his body. The car would still circle. Each time astronaut Runco got the car moving in the track, he would hold the track very carefully, trying not to jar it in any way. But each time, the car would finally slip out of the track because of a tiny motion from his hands. Finally the trick worked and the car stayed in the track as it slowed to a stop. He then released the track, and the car and track floated away.

A student asked what would happen if the track were quickly pulled apart once the car was circling. Would the car jump the loop and continue moving in the track. Mario Runco tried this experiment. The car reached the gap and shot off in a straight line. Once the centripetal force of the track was removed, the car moved away in a straight line. There was no force to make the car continue its circular motion.

Viewers wondered what would happen if the track was released with the car moving inside. Mario Runco demonstrated this several times. Each time he released the track, the car and track would take off together in the direction that the car was moving. The car's motion would cause the track to spin until the car flew away.

SPACE RESULTS: COME-BACK CAN

Under the capable operation of Don McMonagle, the come-back can suffered the same fate as the car. In space, the weight inside the can does not hang down and the can will not roll on a surface. The can depends on rolling friction just like the car does. In space, there is no force to hold the can on a surface. If the can cannot push backward on a surface, then it can't receive the forward reaction force from the surface. In space Don McMonagle could spin the can to wind up the rubber band. When released on a locker, the wound-up can pushed against the locker and the locker pushed the can away. It did not roll along the locker.

Although the come-back can failed to roll, it did provide some very interesting video. First, Don McMonagle swung the can in circles to cause the weight inside to swing around and wind up the rubber band. When he released the can, the weight began to spin inside. As a surprise, the can itself did not turn, but began to wobble in response to the movement of the weight inside. Finally, Pilot McMonagle gave the can a twist and released it. The turning motion quickly subsided because of the motion of the weight inside. Then the can floated away.

8

Space magnets

THREE DIFFERENT SPACE toys use magnets in very different ways. The magnetic wheel has a magnetic shaft that holds it on a metal track. Magnetic marbles have small magnets inside and their behaviors show how magnets interact. Magnetic rings on a rod show how magnetism interacts with other forces. The magnetic wheel and magnetic marbles are available at many toy stores. The magnetic rings can be made from materials purchased at a hardware store.

THE MAGNETIC WHEEL

With the magnetic wheel, you can be a space engineer. Just pretend that this toy is a piece of equipment that works on earth and must now work in space. Your job is to figure out how it works on earth and how it must be modified to work in space. Most NASA hardware, from tools to toilets, must be analyzed this way. Mechanical engineers study how each tool works and how gravity affects its performance. Then they recommend modifications to the tool or new instructions for using it.

Try this procedure with the magnetic wheel. Watch someone else operate the wheel. Notice how they tilt the track to start the wheel spinning along it. Can you think of a way to start the wheel spinning without tilting the track, and to keep it moving without any track tilt. If you can find a way to use the magnetic wheel without gravity's help, then the magnetic wheel could be an effective space toy. If you can't, then the toy has to be modified for space.

Now consider what happens to the wheel as its spinning rate increases. Eventually the magnetic wheel will leave the track. Where does this occur on the track? Does the wheel leave the track at the same place every time? And most importantly, as a space toy engineer, do you think this will happen in microgravity?

A major problem with space tools is keeping them close by. When you set tools down on earth, gravity holds them firmly on the work table. If you release a magnetic wheel, will it remain close by in space? Do you think that the motion of the wheel at the time the astronaut releases the track is important? All of these questions play a fundamental role in designing for space. The problem-solving strategies you use in predicting the magnetic wheel's behavior are much the same as the design challenges of NASA engineers.

MAGNETIC RINGS ON A ROD

This is a toy you can make after a visit to the hardware or electronics store. Many of these stores sell small ring magnets either as utility magnets or refrigerator magnets. The magnets are little disks just under 2 inches (5 centimeters) in diameter with a central hole that is wider than a pencil. The astronauts used six of these ring magnets (three blue and three yellow). The six you buy at the hardware store will probably be gray or silver. After you find the magnets, select a rod to string them on. The rod should be about 1 foot long (30 centimeters) with a diameter just slightly smaller than the diameter of the hole in the magnets. The rod should also be as smooth as possible and not made of iron. Look around the hardware store—a wooden dowel, aluminum tube, or Lucite rod will

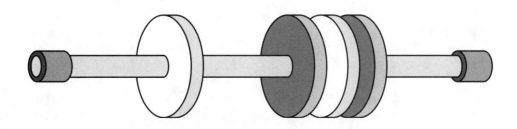

work. You might need to have a longer rod cut to length. The astronauts used a green Lucite rod.

Before assembling your magnetic rings toy, you can experiment with the individual rings. Try hanging two ring magnets on strings. Then bring the strings together. The north pole of one magnet is attracted to the south pole of another magnet. Notice that the poles of the ring magnets are on the flat faces of the rings.

Now stand one ring magnet on its side. Use another ring magnet moving near the first magnet to make it spin away from you. Then use another ring to make the first ring spin toward you. Notice how you use the attraction and repulsion forces of the magnets to make this happen.

With two magnetic rings and the rod, you can also make a chaotic magnetic pendulum. Tape one ring to the end of the rod. Make a tape loop over the other end of the rod so you can swing the rod back and forth. Tape the second ring to a table so that it repels the magnet on the rod. Then hang the rod over the magnet on the table. Pull the rod away and watch as your new pendulum tries to swing. Gravity pulls the swinging pendulum down over the magnet on the table. When the pendulum gets close, the magnets repel. It is very difficult to predict what the pendulum will do after the second or third swing. The interaction between the magnetic repulsion and the gravitational attraction is very difficult to analyze. This is called *chaotic motion*. The swinging behavior is

dependent on exactly how you swing the rod when you release it. Experiment with several different swings. Does the pendulum ever repeat its swing pattern?

Although this is a fascinating earth toy, it will not create a chaotic motion in space. With no downward pulling force, the pendulum will never swing. If you turn the magnet on the table over so that it attracts the magnet on the rod, then you might see one or two swings before the pendulum stops moving directly above the magnet on the table. The magnetic force falls off more quickly with distance than the gravity force does. A magnetic pendulum swinging in space should stop much more quickly than a pendulum pulled by gravity. The magnetic pendulum was an optional activity for the STS 54 astronauts, but was never tried because of time constraints.

To assemble the space toy that the astronauts used, put the six magnets on the rod. Arrange them so that the magnets repel each other. To do this you must have like poles facing each other. Then wrap duct tape around the ends of the rod so that the magnets cannot escape. Your finished space toy should look like the illustration with the poles arranged so that they repel.

Hold the rings vertically and notice the separation between the rings. The repelling force between any two rings is the same, yet the separation distance is greater for the rings nearer the top of the rod. Gravity is the cause of these separation distances. Gravity pulls down on each ring. The magnetic repelling force between the two top magnetic rings is supporting the top ring. The magnetic repelling force between the second and third rings is supporting the top two rings. The magnetic repelling force between the bottom two rings is supporting the top five rings. In this manner, the gravity force pulling the bottom two rings together is greatest and the separation distance is smallest even though the repelling force of the rings is equal between each of the rings.

Hold the rings horizontally and pull the rings together. Then release them while holding onto the rod. Notice how the rings move. The potential energy stored in the magnetic field of the compressed rings turns into the kinetic energy of the moving rings. Each ring pushes the ring next to it which pushes the ring next to it until you reach the end ring. The end ring gets a push from the ring behind it and an additional push as the ring behind it is pushed by the ring behind it . . . and so forth. If you removed the tape on the end of the rod, the end ring would probably fly off.

Using strong tape, attach a piece of string to one end of the rod. Swing the rod around in a circle. Notice how the rings move and what happens to the separation distance between them. Imagine what would happen if your space toy were spun around in circles by an astronaut.

THE MAGNETIC MARBLES

The magnetic wheel stays on a track and the magnetic rings are secured on a rod. In contrast, magnetic marbles are free to move in all directions. The lessons learned from the other magnetic toys can be used to predict interesting things that the magnetic marbles might do in microgravity.

Each of the marbles contains a tiny cylindrical magnet. The magnet's north and south poles cause the marbles to form chains with the north pole of one marble clinging to the south pole of the next. On earth, one marble can lift five or six marbles off a table. The number of lifted marbles indicates the strength of the magnetic force between the marbles as compared with the gravity force.

On a very flat smooth table, marbles roll around in response to magnetic forces. Two marbles rolling toward each other sometimes repel and at other times attract—depending on the orientation of the marbles' poles. Often marbles rolling past one another come together and begin spinning. The magnetic force becomes a centripetal force causing the marbles to stay together as they spin about the point where their poles touch. On earth two circular marble chains can join into one large circle or form a figure eight depending on the direction of their poles.

Both Toys in Space flights carried magnetic marbles. The STS 51D mission had twelve solid-colored marbles of the same size and a thirteenth marble with a piece of thread attached. The STS 54 mission carried sixteen marbles. Each was two-toned with a blue north pole and a yellow south pole. Eight of the marbles were large and eight were small.

Magnetic marbles are available at most toy stores. Look for marbles with seams that you can see. These can usually be separated with a butter knife (a task for adults). Once the solid-colored marbles are separated,

they can be made two-toned with one color for the north pole and the other for the south pole. Use a permanent magnet to identify the poles of the marbles. The blue and yellow marbles carried by the astronauts were assembled from solid-colored magnetic marbles. Try each of these experiments with the marbles, just to get a feel for how they work on earth and might work in space.

Experiment 1: The attraction distance

Place two marbles in a ruler trough. Have opposite poles facing each other. Measure how close the marbles must be before one moves to join the other. Put six marbles in the ruler trough, arranged with opposite poles facing. Pull the marbles apart with three on each end of the trough. Move one set of three toward the other. How close must they come before one chain jumps to join the other?

Experiment 2: The repelling distance

Put six marbles in a ruler trough, arranged with opposite poles facing. Pull three marbles to the end of the ruler trough. Turn them around so the chain is still made up of opposite pole marbles, but when the two chains are brought together two like poles will be facing. Move this chain toward the other. Record the distance at which this chain causes the other chain to move away. Compare this distance with the attraction distance of the three-marble chains in the first experiment.

Experiment 3: Magnetic strength

Holding one marble, pick up another marble and then another. Repeat this procedure until your chain of marbles breaks. Each marble weighs 0.18 ounces (or has a mass of 5 grams). Add the weights of all the marbles that you can pick up with just one marble. This is the total gravity force pulling against the magnetic force between the marble you are holding and the chain of marbles. The magnetic force between two marbles is at least this strong.

Experiment 4: Magnetic marble structures

Try to make the structures shown in the figure. Be certain that the poles of your marbles are marked before you start construction. Once you have a structure built, draw how the poles are arranged. Which of these structures cannot be made?

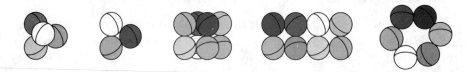

Experiment 5: Marble spins

On a smooth, flat surface, roll two marbles toward each other slowly. Repeat the experiment, but vary the speed and orientation of the poles for both magnetic marbles. Notice how the marbles spin separately and when they join together.

Experiment 6: Marble ring attraction

On a smooth, flat surface, form two rings of six marbles. Push the rings toward each other and watch what happens. Turn over one of the rings and push the rings together again. See if you get the same result.

Experiment 7: Marble dancing

Tape a thread at the north pole of one magnet. Tape another magnet on a table with the north pole facing down. Observe what happens when you hold one marble over the other.

Astronauts tried variations of each of these activities in microgravity. Imagine how the results might be different without gravity holding the magnetic marbles on the table.

SPACE RESULTS: MAGNETIC WHEEL

Getting the magnetic wheel started in space required a completely new approach. With no gravity to pull the wheel down the track, Jeff Hoffman perfected a slinging technique. He jerked the track out from under the magnetic wheel. The wheel's magnetic force kept it holding onto the track. As the track moved past, it began to turn the wheel. Once spinning, the wheel started moving on its own. By timing the track swings, the wheel will move as fast as the gravity-powered earth-based version. Magnetic force kept the wheel from leaving the track as it traveled back and forth.

Two techniques are at work in starting a magnetic wheel in space. First, the wheel's inertia of rest keeps it still as the track is pulled away from it. The magnetic force causes the wheel to hang onto the track and to start spinning as the track moves. Once the wheel begins spinning, it will continue spinning for a moment until the operator pulls the track away from the wheel again. The second spinning technique involves a circular motion of the track. Jeff Hoffman put the wheel next to his hand

and swung the track around in a broad arc. The wheel moved toward the end of the track. Its spin then carried it back inward along the track. Once the wheel returned, another swing sent it flying outward again. At each swing the wheel's speed increased.

In suggesting this toy, experimenters had expected an unusual behavior when the toy was released. Perhaps the spinning wheel would retain its spin axis and the track might move around the wheel. Jeff Hoffman first held onto the track and let friction cause the wheel to stop. It came to rest at the top of the loop in a position that could occur only in space. Then he carefully released the track as the wheel was moving away from him. He gave no push to the track, yet the track moved away as if it were moving on its own. To explain this behavior, you must think about the magnetic wheel's total momentum. The wheel was moving away from Jeff Hoffman. When the track was released, the outward-moving wheel dragged it along. When the wheel reached the track's turn, it continued to move forward and thereby caused the track to turn. As it slowed down, the wheel continued to move along the track, causing the track to turn. The magnetic wheel moved in the direction that conserved momentum.

Since the wheel was spinning, angular momentum also had to be conserved as the wheel slowed down. This was accomplished by the slow turning of the track. Because the track is much bigger, its slight turning has the same angular momentum as a faster spin of the wheel. As the magnetic wheel floated away, its behavior was determined by these momentum conservation principles. Once Jeff Hoffman gave the track a slight push before releasing it. The result was a much more violent tumbling behavior. A little turning of the more massive and larger track gave the magnetic wheel system a great deal of extra angular momentum.

Perhaps the magnetic wheel could be modified to increase its gyroscopic stability. If the wheel were more massive and if the mass were distributed along its rim, then the wheel would have more angular momentum for its spin. When released, the track would have to spin more around the wheel as the wheel slowed down.

SPACE RESULTS: MAGNETIC RINGS ON A ROD

John Casper chose the magnetic rings on a rod as one of his space toys. First he pushed the rings to one end of the rod and released them. The rings quickly separated evenly along the rod. Every other ring seemed to vibrate for a moment before the rings stopped moving. He then pulled two rings together on one end and released them. The compression moved along the rod from ring to ring. Although much less strong, this

wave behaved like the wave produced in a coiled spring. These rings are "connected" by the magnetic field between them in much the same way as the coils of a spring are physically connected. Energy can be stored in the compressed coils of a spring and in the magnetic field of the rings when the rings are pushed together.

John Casper then turned the rod vertically and squeezed the rings together again. Once again the rings spread apart. Although this was the same behavior as before, it seemed very strange to see the rings spread evenly apart on a vertical rod.

Finally John Casper swung the rod like a baton twirled by a drum major in a marching band. First he twirled it slowly and nothing seemed to happen. Then he increased the twirling rate and the rings began to move outward. The magnetic force that held the rings apart could not stop the outward momentum of the rings as each tried to move in a straight line. Three rings settled at each end of the rod. The spacing between the outer two rings on each end was much smaller than the spacing between the inner two rings. In effect, the magnetic force between the rings became the centripetal force pushing the inner rings toward the center of the spinning rod.

SPACE RESULTS: MAGNETIC MARBLES

Earth-based marble activities pale in comparison with marble behaviors in weightless freedom. In space, magnetic effects dominate marble motions in all three dimensions. The rolling friction caused by gravity disappeared in space. On earth, this friction is too great, even on the smoothest table, for many of the most exciting marble maneuvers.

Jeff Hoffman's marble experiments produced fascinating results. Two space marbles aimed toward one another stuck together and spun as a pair. Marbles had to come close enough for the magnetic forces to overcome the marbles' forward momentum. A third marble thrown toward the other two joined to form a chain with opposite poles touching. The third marble's momentum produced a drift in the resulting three-marble chain as well as a spinning motion for the chain. The fourth marble caused similar behaviors in the growing marble chain. Each marble attaching to the end of the chain added angular momentum to the chain. The chain began spinning around its center—the junction of the two interior marbles. The fifth marble actually moved around the chain to attach on the proper end. The sixth marble caused the chain to begin wobbling, with the center and ends swinging in opposite directions. The seventh marble increased the wobbling motion as it contributed its momentum to one end of the chain. Soon the two ends came close enough for the magnetic force to pull them together to form a circle. The eighth and ninth marbles veered toward the circle and were absorbed as the circle's diameter grew. As each marble hit, its momentum caused a spinning motion of the marble circle. The tenth marble attached in a stable position between two marbles in the circle. The eleventh marble joined the tenth and gave the circle a tail. The twelfth marble then attached to the other side. The joining behavior of each marble depended on its motion and the orientation of its magnetic poles.

Jeff Hoffman formed the twelve marbles into two circles. On earth he had discovered that the six-marble circle was very stable and could be tossed in the air without breaking apart. As in earth-based experiments, the behavior of the magnetic circles depended on the magnetic orientation of each circle. The two interacting marble circles produced three possible results. The two circles could repel, attach to form a figure eight, or merge into a large circle. All three behaviors occurred in space.

Greg Harbaugh replicated Jeff Hoffman's experiments combining marble rings. His marble rings each had eight marbles. One ring was made of large marbles and the other ring was made of small marbles. He could get the two rings to repel each other and to attract each other forming a figure eight. The figure eight would not form into a single chain as Jeff Hoffman's marbles had done. The difference in the sizes of the marbles is the probable reason for this.

Both Jeff Hoffman and Greg Harbaugh experimented with swinging a chain of magnetic marbles. The magnetic force holding the marbles together became the necessary centripetal force to keep the marbles moving in a circle. As the spinning rate increased, the required centripetal force increased. When the needed centripetal force became greater than the magnetic force holding the chain together, the chain broke, and the marbles flew off in a group traveling away along a straight-line path. The inertia principle led to this marble chain breakdown. In all three attempts, the break came between the first and second marbles in the chain. Students familiar with playing "crack the whip" thought that the last marble would be the first to break away. But in a marble chain, the innermost link is pulling in on all the other marbles. When the outward inertial force overcame the magnetic pull, all of the marbles flew off at this first link, except for the one attached to the string.

Jeff Hoffman kept the marbles in a pocket during the experiments and carefully brought out each one as needed. Along with the yo-yo and car, these tiny toys became his favorites. Their motion in weightlessness dramatically illustrated inertial and magnetic forces operating without the interference of gravity and friction. Single marbles remained in straight-line motion, as inertia dictates, until a magnetic force changed the motion. The momentum of the drifting marbles was always conserved during a collision. Jeff Hoffman commented that he could have played with the marbles all day, or at least until Commander Bobko told him to get back to work.

Greg Harbaugh on the STS 54 mission tried to improve on some of Jeff Hoffman's results. Greg Harbaugh's marbles were blue and yellow so everyone could tell the north end from the south end of each marble. He placed two marbles in the air. At a distance of approximately 1 foot (30 centimeters), the attraction between opposite poles was strong enough for the marbles to float together. If like poles faced each other, the marbles would repel.

Greg Harbaugh's two-toned marbles demonstrated elegant spinning motions. When one marble was moved close to another, it would make the other floating marble spin. Each movement of the first marble past the floating marble increased this spinning motion. The north pole of the first marble pulled on the south pole of the floating marble as it passed, giving that part of the floating marble a tug.

Greg Harbaugh also produced the spinning behavior that Jeff Hoffman observed when two marbles joined. Because the STS 54 marbles were two-toned, he also noticed that the marbles rocked back and forth as if their magnets were trying to align themselves. This same rocking motion repeated when one marble joined two others. Although the threesome turned around the center marble, the chain also seemed to rock. Next, Greg Harbaugh took two chains of four marbles. Depending on how their poles were facing, he could get them to join into a single chain or to repel one another.

In another marble experiment, he arranged a target of four marbles—three small ones and one large one. The large marble had one small marble on one side and two small marbles on the other. Then he threw a small marble toward this chain so that it would join up on the side with the single marble. The resulting chain began to spin in the direction that the final marble was moving while circling around the larger marble in the center. This chain also showed an unusual wiggling motion.

In every experiment, Greg Harbaugh captured the predicted motion along with a strange rocking or wobbling of the marbles. The only explanation given thus far for this motion is the interaction between the marbles and the magnetic field of the earth. The earth's weak field does not appear to affect the marbles as they roll on a terrestrial table. The friction between the marbles and the table is strong enough to dampen any force trying to turn the marbles. But in space, the marbles can respond to the presence of the earth's weak magnetic field. Most onboard camera shots show the yellow south pole of each marble, indicating that each marble was aligned with the magnetic poles of the earth.

9

Space vibrations

THREE DIFFERENT TOYS operate by vibrations: the coiled spring and paddleball of the STS 51D mission and the coiled spring and Jacob's ladder from STS 54. The paddleball uses an elastic string stretched between a ball and a paddle. The coiled spring can be stretched and released to produce wave motions. The Jacob's ladder depends on gravity on earth. See if you can predict how it generates a stretched vibrating motion in space.

THE JACOB'S LADDER

If you don't have a Jacob's ladder, you can begin by making your own. All you need are two decks of playing cards or some wood blocks, some wide ribbon, and tape. Then you can discover the special way the Jacob's ladder works on earth and imagine what it might do in space.

You will need

- Three lengths of wide gift wrap ribbon—each at least 18 inches (45 cm) long
- Six pieces of wood—about 1.5 inches by 3 inches by 0.25 inch (about 3.75 cm by 7.5 cm by 0.625 cm) (or two decks of regular playing cards)
- good quality clear tape
- scissors
- stapler with staples

1. If you are using wood blocks, sand them until they have smooth edges and then paint them in bright colors. Wait until the paint dries.

If you are using two decks of playing cards, divide the decks into six piles of seventeen cards each. Throw the remaining two cards away. These piles of cards will be the blocks in your Jacob's ladder. Use tape to cover the ends of the card piles. Make the tape smooth.

2. When ready, place all six blocks (or piles of cards) in a long row on a flat table. Then add the ribbons, one by one.

—Center ribbon

A. Fold the ribbon over and staple it to the far center edge of the first block.
B. Lay the ribbon over the top of the first block.
C. Place the second block on top of the ribbon and staple the ribbon to the block at the edge closest to the first block.
D. Place the third block under the middle ribbon. Staple this block to the ribbon on the end closest to the second block.
E. Continue this procedure until you reach the sixth block. Then staple the ribbon to the end closest to the fifth block. The middle ribbon does not stretch over the sixth block.

—Two outer ribbons

A. Staple these ribbons to the edge of the first block that is closest to the second block.
B. Lay these ribbons over the second block.
C. Staple these ribbons to the end of the second block that is closest to the third block.
D. Place both outer ribbons under the third block and staple at the end closest to the fourth block.
E. Continue until you reach the sixth block. Then run these two ribbons over the block and attach on the far end of the row of blocks.

3. Your Jacob's ladder is complete. Hold it up and tilt the top block over. This should start a chain reaction all the way down the ladder. You might have to adjust the tightness of your ribbons to make your ladder fall smoothly. Notice the location of the ribbons when the blocks flip. What force makes the blocks flip?

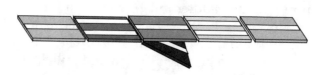

As another experiment, hold the Jacob's ladder in a level horizontal position with your hands on the end blocks. Pull the end blocks apart with tension. Then use your hand to fold the end blocks up or down to make changes in the ladder. Pull the end blocks apart and notice what happens. Repeat this activity until the blocks flip up or down in different positions each time you stretch the ladder apart. Use these activities to predict how the ladder might behave in space.

THE PADDLEBALL

The paddleball and coiled spring illustrate two extremes of elastic stretch toys—each different from the Jacob's ladder. Although the elastic string in the paddleball brings the ball back, the ball's exact path and final position are different for each paddle stroke. A coiled spring is also stretched, but its behavior is predictable on earth and in zero gravity. Stored spring energy brings the coiled spring back into its coiled position on earth and in space.

Hitting the paddleball is a challenge for most earthlings of any age. Beginners are first successful at batting the ball straight down. In this direction, gravity becomes a correcting force. If you hit the ball downward at an angle, gravity will make the ball's path more vertical. The ball's return location from a downward hit is almost always vertical and easy to anticipate. For short paddlers, the ball might hit the floor, bounce away at an angle, and cause great paddling confusion. To solve this problem, a beginning paddler can hit the ball forward and downward, and then allow the ball to swing under his arm upon its return. After swinging backward, the ball will swing forward again. When it approaches the paddle for a second time, the paddler hits it and repeats the ball motion. This extra swing gives valuable reflex time. The technique takes advantage of the natural slowing of the ball as it rises against gravity's pull. For a downward paddler, the gravity and elastic effects help even out the ball's speed. As the gravity force accelerates the downward moving ball, the stretching string acts to slow it down.

Each time the paddle hits the ball, the impact force changes the ball's direction of motion. The action force of the ball moving forward produces a reaction force as the paddle tries to move backward.

The truly gifted paddleball player plays with the paddle angled upward. In this position, the paddler must strike with force to compensate for gravity's downward pull. If the paddling speed is too slow, the returning ball will drop too much to be hit again. The elastic return force must be much stronger than gravity's downward pull. Instead of being a stabilizing force, gravity makes any small misguiding of the ball much

more disastrous. In the upward version, the ball is also traveling its fastest at the time when you must hit it with the paddle.

To demonstrate paddleball dynamics on earth, stretch apart the ball and paddle. If the ball is pulled downward, it will return to the paddle. If it is stretched upward, it must be dropped directly over the paddle to hit correctly when released. If stretched to the side, gravity will cause the ball to drop below the paddle on its return.

On earth a paddleball stores two forms of potential energy. As the elastic string stretches out between the paddle and ball, elastic potential energy increases. When released, the ball rushes toward the paddle, converting this potential energy into kinetic energy. This simple motion is confused by gravity. Gravity makes the ball go faster as it falls and causes it to slow down as it rises. Hitting a paddleball downward is very different from hitting a paddleball upward. In space the gravity effect disappears. What effect will this have on paddling in any direction?

THE COILED SPRING

The stretched spring's motion is controlled by the strength of its spring. The coiled spring's most famous earth-based behavior is "walking" down stairs. The motion begins as a coiled spring is stretched between two adjacent steps with its end coils placed on each step. Energy stored in the stretched coiled spring pulls the coiled spring together. A gentle push starts the coiled spring moving forward. Gravity increases the coiled spring's speed as the coils from the upper step begin falling down to the next step. As the last coils fall onto the others, the coiled spring has enough forward momentum for the top coils to bounce off the coiled spring and onto the next step. Now the process can repeat itself all the way down the stairs. Although gravity directs the walking of the coiled spring, it is the restoring force of the coiled spring that drives the motion.

A coiled spring stores energy as you stretch it apart. When released, the coiled spring rushes together trading the mechanical potential energy stored in its spring for the kinetic energy of motion. Kinetic energy can also be sent along a stretched coiled spring. Imagine a coiled spring extended between two astronauts in space or two earthlings sitting on the floor. If one end of the coiled spring is vibrated, the vibration will travel along the coiled spring to the other end. A strong vibration can even bounce off the second person's hand and travel back along the coils. In this way the coiled spring carries energy from one person to another. Because the ends of the coiled spring are held stationary, the coiled spring itself does not do any moving as it carries the wave energy.

The coiled spring is a classic demonstrator of wave motions on earth and in space. In any wave motion, a disturbance travels through space or through a substance without causing any permanent motion in the substance.

There are two basic kinds of wave motion: longitudinal waves and transverse waves. Sound waves are longitudinal waves. The stretched spring readily makes longitudinal waves. To demonstrate longitudinal

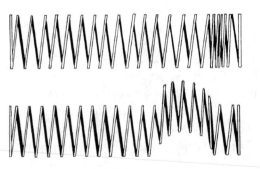

Longitudinal or compression wave

Transverse wave

waves, stretch a spring between two people. One person squeezes several spring coils together. The place where the coils are close together is called a *compression*. When the compression is released, it moves along the coiled spring toward the other person. This is a longitudinal or compression wave. If the person at the receiving end of the compression wave holds the end coil very still, the compression will "bounce" at the spring's end and move back along the spring toward the person who created it. The coiled spring transmits a wave and produces an echo, just as a sound wave bouncing off a canyon wall echoes the speaker's voice.

A sound wave has three properties illustrated by the coiled spring. It has speed as it travels along the spring. This speed is determined by the springiness of the coiled spring rather than by any behavior of the person sending the wave. To change the wave's speed, you can change the amount of stretch in your coiled spring or exchange your coiled spring for one with a different amount of spring. A brass coiled spring, for instance, is much looser than its steel equivalent. In like manner, the speed of sound in air at sea level at a temperature of 68°F (20°C) is 769 miles per hour (344 meters per second). Yelling loudly will not increase the speed of the sound wave you produce. Scientists and pilots often refer to the speed of sound at sea level as Mach 1. A plane breaks the sound barrier and produces a sonic boom as its speed exceeds Mach 1. In less than one minute, the shuttle reached this speed. A minute later the shuttle's speed attained Mach 4.5. When the shuttle began orbiting, its speed was Mach 26.

By increasing the number of coils in the compression wave, you create a stronger compression. A stronger compression will last longer as it moves back and forth along the coiled spring. Yelling makes a stronger compression in a sound wave. The vibration in the listener's eardrum is stronger and the person hears a louder sound. The louder sound produces a stronger echo on canyon walls.

You can also change the number of compressions traveling along the coiled spring each second. This number is the wave's *frequency*. Changing the frequency of a sound wave changes the sound's pitch. Many waves passing by a point each second indicate a high frequency and are analogous to a high-pitch squeaky sound wave. A few waves moving along the coiled spring are like a low-pitch sound.

You can also use a coiled spring to illustrate special wave properties. If your timing is good, you can create standing compression waves, where certain parts of the spring are always compressed and other parts are always extended. The compressed areas are called *nodes*. Try making standing waves with different numbers of nodes along the spring. You might need a partner pushing in and out on the stretched spring.

Jeff Hoffman and Rhea Seddon used the sideways motion of a stretched coiled spring to produce another kind of wave—a transverse wave. In this wave, the coils move from side to side as the wave moves forward along the coiled spring. Compare this behavior with the longitudinal wave where the coils move forward and backward as the wave moves forward along the coiled spring. A transverse wave will also bounce and return. The holder of the receiving end of the coiled spring should make the end coil very rigid. The incoming wave will bounce off. If the wave comes in moving to the right, it will bounce off to the left as it travels back along the spring. Transverse water waves in a pan or on a lake behave in this manner.

You can also make standing transverse waves. First, stretch the spring between yourself and a friend. Place it on a hard smooth floor or table. Move the spring back and forth quickly to generate sideways waves traveling along the spring. Now try to move the spring back and forth so that a standing transverse wave is created. In this standing wave, there will also be places called nodes where the coiled spring does not move and the coiled spring will go back and forth on either side of the nodes. Try making standing waves with different numbers of nodes.

Light waves are transverse waves. Like sound waves, light waves can be stronger or weaker and can have higher or lower frequencies. Light waves, however, travel through the vacuum of space. They do not need an atmosphere or any other medium to travel from place to place.

Like all transverse waves, the coiled spring version has a wavelength, a frequency, and an amplitude. When describing a coiled spring wave, the motion to one side is called a *crest* and the motion to the other side is a *trough*. These words make it easier to describe the coiled spring's wave properties. The highest part of a water wave, for instance, is the crest. The dip between crests is the trough.

Amplitude is the distance from the wave crest straight down to the rest position of the coiled spring. Higher amplitudes indicate stronger waves. Stronger waves move back and forth along the coiled spring for a longer time before dying out. The coiled-spring operator must use more muscle energy to produce them. Like the longitudinal sound wave, the transverse wave moves through the water without carrying the water forward. The transverse coiled-spring wave does not cause the coiled spring to collect in the receiver's hand.

The *frequency* of a transverse wave describes the number of wave crests passing along a point on the coiled spring in one second. The *wavelength* is the distance between two consecutive crests. A coiled-spring wave that goes up and down only once has a wavelength equal to the distance between the ends of the coiled spring. If two complete waves fit exactly in the coiled spring, then the wavelength is half the distance between the ends of the coiled spring.

There is a third kind of wave you can make with a coiled spring called a *torsional wave*. Stretch the spring out on a hard smooth floor or table. Place pieces of tape on the tops of eight loops of the spring, evenly spaced along the spring. Twist one end of the spring and watch the twist move along the spring. The pieces of tape should make it easier to see the twisting. Compare the speed of this torsional wave with the transverse and compression waves. Notice how much less the spring actually moves and how much faster the waves are.

SPACE RESULTS: JACOB'S LADDER

Susan Helms discovered that the Jacob's ladder makes a beautiful space toy. It does not flip over like it does on earth because there is no force to pull the loose block downward. Instead the blocks behave like a loose slinky. When she pushed the blocks together, they vibrated in and out like an accordion playing by itself. When the blocks hit each other, the impact pushed them back apart. When they were stretched out, the ribbons pulled them back together again.

Susan Helms also stretched the Jacob's ladder between her hands and then pushed it together. When she pulled it apart, one block would be sticking up or down. Then she pushed it together again. The next time she pulled the ladder apart, another block would be sticking up or down. There seemed to be no pattern to which block would stick out of the ladder when it was stretched. The loose ladder floated around like a snake wiggling back and forth. The strength of the vibrations depended on the force of Susan Helms' hand as she released the ladder.

SPACE RESULTS: PADDLEBALL

In general, paddleball playing is easier in space because orientation is not a factor. Don Williams could strike the ball at any angle and at almost any speed. The elastic would always pull the ball back toward the paddle. Slower, more elegant paddling was definitely possible. The speed and the required reaction time depended only on the paddling force, not on the paddling direction.

Students asked if the paddleball would paddle by itself in space. This trick required advance planning about when to release the paddle. As the stretched elastic pulls together, the ball moves back toward the paddle. Upon release, both the ball and the paddle move, but the lighter ball moves much more. The only instance of self-paddling occurred when Don Williams released the paddle as the ball was coming back. After one hit, the paddle began to tumble. In tumbling, the paddleball paddle dives

head-first toward the ball. The pull of the stretched elastic string produces this tumbling complication. The string is not attached at the best place on the paddle. To make earth-based paddling easiest, the elastic is stapled to the center of the paddle. For self-paddling in space, it should be attached at the center of the paddle's mass, located closer to the paddle's handle. A redesigned paddleball paddle with a handle behind the round face might work as a toy in space and be capable of self-paddling for a few hits.

The paddleball's ball is small and light so that the earth-based operator can hit it gently and get a returning ball without breaking the elastic string. In space a larger more massive ball could be used. For a given paddling force, a ball with greater mass will move more slowly, and its larger size would be easier to hit.

SPACE RESULTS: COILED SPRING

Both Rhea Seddon and Jeff Hoffman worked together demonstrating the coiled spring's wave behavior. Rhea Seddon chose the coiled spring because she had enjoyed it as a child. Long before the Toys in Space project, Jeff Hoffman had wanted to take a coiled spring into space.

The coiled spring proved to be an excellent wave demonstrator in microgravity. The Hoffman and Seddon team created transverse standing waves with one and then two nodes along the coiled spring. They also caused standing compression waves along this transverse wave. In the STS 54 mission, Mario Runco played with the coiled spring. He was able to create standing waves along the spring stretched between his hands. Although nothing really unusual happened with the coiled spring in space, its waves are some of the most beautiful science demonstration footage from the missions.

Coiled spring space-walking met with instant failure. When the coiled spring was stretched between two space steps, it floated off and recoiled. Only a slight rebound vibration indicated that the coiled floating spring had just been stretched. Depending on the direction of stretch and release, the coiled spring can acquire forward momentum and begin drifting across the cabin. The unevenness of the impact can also cause the spring to rotate.

The spring can be rocked from hand to hand with the end-coils resting in each hand. On earth this simple behavior results in closely packed coils resting in each hand. In zero gravity the coiled spring had to be tossed to move it back and forth between hands. The coils remained more evenly distributed along the coiled spring. When one hand released the coiled spring, the coils came together in the other hand and then bounced

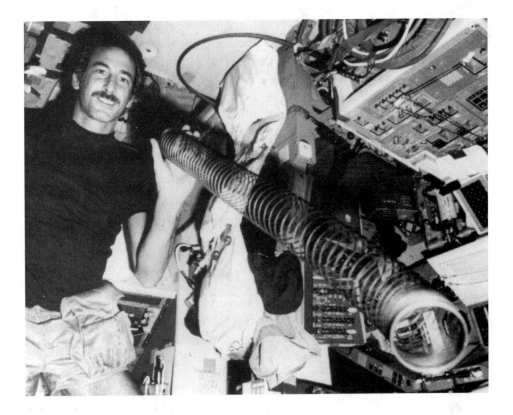

away for another smaller extension. After several more bounces, the coiled spring returned to its compact state.

Jeff Hoffman tried spinning the coiled spring between his hands and then tossing it. While spinning rapidly, the coiled spring remained stretched out. The spinning force acted like a centripetal force keeping the coiled spring from flying apart. When the spinning slowed, the coiled spring immediately returned to its closed position. Although interesting, this behavior was too short-lived to be an impressive demonstration. The worst toy disaster occurred when this spinning coiled spring "dumbbell" flew across the string of Dave Grigg's moving yo-yo. Soon the coiled spring and yo-yo were thoroughly tangled.

Jeff Hoffman was much more successful when he began "yo-yoing" with the coiled spring. He threw the coiled spring outward just as you would release a yo-yo. The weightless coils flew forward until the spring force overcame their outward momentum and pulled the coiled spring back together in his hand. This became his favorite way to play with a coiled spring. Jeff Hoffman was most impressed with the symmetry of the coiled spring's behavior as it drifted outward, slowed down in its motion, stopped, reversed direction, and began to return. As it approached his hand, its speed reached the tossing value. By letting the coiled spring fly past his hand, he could watch the coiled spring repeat this behavior behind his head.

Like the spinning coiled spring, the yo-yoed coiled spring ventured into the territory of other toys. In this instance, the problem was the jacks. Moving spring coils stretched through Rhea Seddon's floating jacks, sending jacks flying in all directions. The light jacks were no momentum match for a sailing spring.

10

Spinning space toys

SPINNING TOYS are naturals for space flight. On earth, gravity causes spinning toys to run down and tip over. In space, a spinning toy can do its spinning without any interference from outside forces. The STS 51D astronauts chose the gyroscope and top. The STS 54 crew added a gravitron, tippy top, rattleback, and spinning book. The spinning book is a toy you can make from a small thin book, and it provides a surprise twist to your investigation of spinning objects.

THE TOP

The simplest of all spinning toys is the top, and each crew had one top on board. Colonel Bobko of the STS 51D mission chose a metal push top. A simple downward push of the handle starts the top spinning. As the top spins, it slows down. As it spins more slowly, it loses some of the angular momentum that keeps it upright. Gravity causes the top to wobble (or precess). As the spin rate drops, the wobbling increases.

Since gravity causes the wobble, it is easy to predict that the top will not wobble in space. But is gravity required to operate the top? Will it be as simple to spin the top in space as it is on earth?

The STS 54 top had an entirely different design. The top is usually called a "tippy top" or "magic top." When you spin the handle, the top

immediately begins tipping over. If your spin is fast enough, the top will end up upside down, spinning on its handle.

It's the tippy top's design that causes this unusual behavior. Its rounded bottom has no point for the top to spin on. As the top spins, gravity pulls it over and the axis wobbles around the smooth bottom. Finally the top flips onto the handle, which defines a stable axis. Gravity seems to be the cause of the tippy top's behavior. If it is, then the top should not tip over in microgravity.

THE SPINNING BOOK

Any hardbound book can become a spinning toy. Choose a book that is long and thin. Put rubber bands around your book to keep it closed. There are three different ways you can spin a book:

1. Longways. Hold the book's spine toward you in the vertical position and spin the book with both hands. Does the book spin or does it wobble?
2. Shortways. Hold the book's spine toward you in the horizontal position. With your hands at each end, spin the book. Does it spin or wobble?
3. Frontways. Hold the book facing you. Place one hand on the spine and the other on the side where the book opens. Spin the book. Does it spin or wobble?

On earth, the book spins in the first two modes and wobbles in the third. As described above, there are three axes of spin: the long axis parallel to the binding, the short axis in the plane of the binding, and the

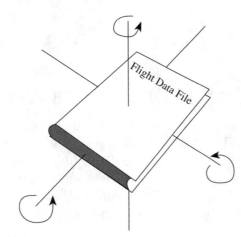

medium axis perpendicular to the binding. All three axes run through the center of mass of the book. The long axis spin has the smallest moment of inertia with most of the mass near the axis. The short axis spin has the largest moment of inertia with most of the mass at the greatest distance from the spin axis. Both the short axis and long axis spins are stable. The medium axis is unstable and the book immediately begins to wobble.

THE RATTLEBACK OR SPOON

The rattleback (also called a celt or space pet) is the most unusual of spinning toys. It spins easily in one direction on a smooth surface. When you spin it in the other direction, it slows down, stops, and starts spinning in the first direction. Look in a museum or science center gift shop for one of these unique toys or make your own with an inexpensive metal spoon.

Bend the spoon handle over the spoon. The handle must be bent so that the spoon still balances on its bottom and doesn't turn over when you spin it. The handle must also cross the spoon at an angle, not directly along the long axis of the spoon. Your finished bent spoon should look like the one shown in the illustration.

When you spin the spoon, it will behave like the rattleback with a stable spin in one direction and a wobbling motion in the other. Be sure to spin your spoon on a smooth surface to get the full effect. On earth, the rattleback or spoon will stop and reverse its direction of spin. Is this a property of the object's spinning motion, or is it caused by gravity? Do you think the rattleback will spin in either direction in microgravity?

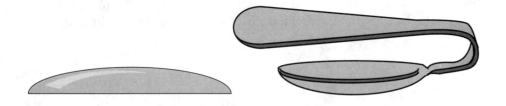

THE GYROSCOPE AND GRAVITRON

To predict what a gyroscope or gravitron will do, you need to experiment with them on earth. The *gyroscope* is a top surrounded by a cage. You can hold onto the cage and try tilting and twisting the spinning gyroscope wheel. A *gravitron* is a gyroscope with the wheel completely within a plastic casing. With the spinning wheel protected, you can tilt and balance a gravitron without grabbing the wheel and stopping the spinning motion.

If you have a gyroscope or gravitron, try these experiments to discover how spinning objects behave. First, start the toy spinning. Then try to change its spinning. Spinning objects turn around an axis. The spinning object pushes against your hand when you try to change this axis.

Start your gyroscope or gravitron spinning rapidly. Immediately place it on its stand and watch how long it spins before it falls off the stand. Place a nonspinning gyroscope on the stand and watch what happens. What causes the gyroscope to stay on the stand? What causes it to fall off the stand? How does the gyroscope behave as it slows down?

Start your gyroscope spinning and set it on a string. Tilt the string and see if you can make it walk like a tightrope walker. Spin the gyroscope again, set it on a table, and push it with a string or cord. Try to make the gyroscope walk along the string.

Tie a string on one end of a gyroscope. Swing the gyroscope around by the string. Notice how the nonspinning gyroscope orients its axis. Now start it spinning and swing it around by the string in circles. Notice how the gyroscope orients itself to keep its spin axis stable.

In a familiar demonstration, a student sits on a swiveling chair and holds a spinning bicycle wheel by handles attached to its axle. As the student tilts the turning wheel, the wheel resists the turning motion by pushing against the student. The student begins spinning around. If he tilts the wheel in the opposite direction, he will spin the other way. If a free-floating astronaut were comparable in mass to his gyroscope, he

would have the same experience as the student. An astronaut's attempts to turn a gyroscope in space would cause him to turn around in the cabin.

The gyroscope and top are both flywheels storing energy in their spins. A spinning object stores more energy and has more angular momentum if its mass is farther from the spin axis. The top's wide body design puts much of its mass far from the spin axis. In like manner, it requires much more energy for an ice skater to spin quickly with her arms outstretched. When the skater pulls in her arms, she will spin more rapidly to conserve angular momentum.

A nonspinning gyroscope tumbles when tossed. Once spinning, the gyroscope moves gracefully in the direction that it is pushed. The terrestrial football suffers the same fate. For maximum speed and catching ease, it must fly nose-first through the air. To stabilize the football in flight, the quarterback spins it, resulting in gyroscopic stability. A child's first frustrating football toss tumbling end over end shows the need for spin stability.

THE YO-YO

Watch a yo-yo in action—moving down and then back up the string. Notice how its speed changes. Is gravity controlling the motion of the yo-yo, or is the conservation of angular momentum what keeps the yo-yo spinning along the string? On earth, it is very difficult to sort out these different effects. Before the STS 51D mission, many students and scientists thought that the yo-yo would be impossible to operate in microgravity.

Dave Griggs chose the yo-yo because he liked to perform yo-yo tricks. It was most important for the astronaut using the yo-yo to be able to perform each trick on earth. Otherwise we would never know if the floating

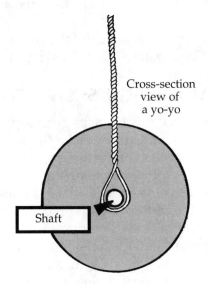

Cross-section
view of
a yo-yo

Shaft

environment made the trick impossible or if the astronaut did not know how to do the trick. Dave Griggs practiced and perfected the sleeping yo-yo, shoot the moon, the flying saucer, and around the world.

Dave Griggs chose an inexpensive toy store yo-yo. There was nothing special about this yo-yo that made tricks possible, except that it was a sleeper yo-yo with the string looped around the shaft. Only a sleeper yo-yo can stay at the bottom of the string. For those who have not held a yo-yo in many years, here is a brief description of each trick. Imagine trying to do each one in microgravity.

To sleep a yo-yo, you send the yo-yo down the string and hold the string lose as the yo-yo reaches the bottom. The yo-yo spins around the shaft (or sleeps) until you jerk on the string and bring the yo-yo back up to your hand.

To walk the dog, you guide a sleeping yo-yo along the floor and attempt to walk it as you would a dog. Then you return it to your hand while the yo-yo still has enough spin to climb.

To rock the baby, you start a yo-yo sleeping at the end of the string. Quickly grab the string and hold it around the yo-yo so it looks like you are rocking a baby. This trick depends on the yo-yo sleeping for at least 10 seconds.

To perform around the world you send the yo-yo out the string and let it sleep at the end. Then quickly swing the yo-yo around in a complete circle and pull the yo-yo back to your hand.

To shoot the moon you send the yo-yo out in front of you. When it returns to your hand, you swing it around your wrist and aim it upward toward the ceiling.

To perform the flying saucer you send the yo-yo out at a sideways angle and try to bring it back before the string tangles.

SPACE RESULTS: TWO TOPS

The metal push top behaved much like its gyroscope cousin. Without any spin, it tumbled like a tossed ball. To make the top spin proved to be a more difficult task in space than on earth. Commander Bobko pushed the top handle down and the top began spinning. He pulled back up on the handle and the whole top came up in his hand. Pumping the top required two hands, one to push and the other to hold down the top. Once spinning, the top bounced off the table and drifted across the cabin like a flying saucer. A gentle push moved the top sideways with no tumbling. Since it did not rub on a table, the top spun much longer than its earth-based counterpart. Like the gyroscope, the slowing top stayed vertical with no wobble. Although not as well balanced as the gyroscope, the spinning top showed that gyroscopic stability is a property of all spinning toys.

The tippy top received its first flight experience on the KC-135 in 20 seconds of freefall. Without a downward pull, the top floated. The operator pushed the top against the floor of the airplane to start it spinning. Then it drifted spinning across the airplane. There was no evidence of the tippy behavior. Without a surface to rub against and a force pulling downward, the tippy top behaved like any other spinning object. It kept a stable axis of spin and did not tip or flip.

SPACE RESULTS: RATTLEBACK AND SPINNING BOOK

As predicted, the rattleback had no trouble spinning freely in either traditional direction. Gravity is, in fact, the cause of the wobble of the rattleback on earth. In space the wobbling motion does not occur. The rattleback would also spin well in its long axis, which was like spinning a football, and short axes, which produced the traditional spinning motion and an end-over-end motion.

The spinning book was not tried in space. The klackers were used instead. First Mario Runco taped the klackers open. Now the klackers have three axes of very different lengths and mass distributions. The short axis would be like spinning the klackers when they are lying flat on a table. This axis is probably stable, but was never tried in space. The long axis runs from klacker ball to klacker ball. First Mario Runco held a klacker ball in each hand and gave the toy a spin. The result was an elegant stable motion of the klacker handle spinning end-over-end.

Finally, Mario Runco tried to spin the outstretched klackers around the handle. This is the middle axis and the one that should be unstable. The result was dramatic. First the klackers started to spin around the handle. Then the whole toy dipped and started spinning around the long axis, then the toy dipped again and the axis shifted. This unstable dipping and turning motion continued until Mario Runco rescued the toy from its confusion. Undoubtedly the middle axis was very unstable.

SPACE RESULTS: GYROSCOPE

Spinning toys win any space predictability contest. Governed by their spinning motion, they performed in space like they do on earth, only with more elegance and style. Once a spinning toy starts spinning, its motions tend to conserve angular momentum. In simple terms, the toy resists any attempt to change the tilt of its spin axis.

The gyroscope is the fastest of the spinning toys. Its behavior was the most indicative of spinning toy motions in space. Commander Bobko first pushed his nonspinning gyroscope and watched it tumble through the middeck. He then wound up the gyroscope and gave the string a sharp tug. The spinning began. As the gyroscope's spin axle rubbed against the cage, the cage also began to spin. No amount of gentle nudging or pushing could make this gyroscope change the direction of its spin axis. With each push it drifted upright through the cabin.

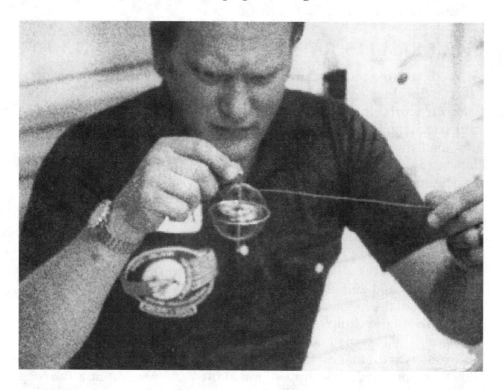

Rubbing between the gyroscope and its cage and air resistance caused the gyroscope to very gradually lose its spin. The slow-down rate was much less than on earth. On earth, gravity causes the slowing gyroscope to wobble or precess. To produce this motion, the falling gyroscope's spin carries its axis sideways. In this manner, its spin axis traces out a circle. Once this precession starts, the earth-based gyroscope experiences more axle friction that further slows the spinning rate. The precession circle widens and the gyroscope topples over.

On earth you can balance a spinning gyroscope on a string or a small stand. You can touch its cage with a string causing the gyroscope to move along the string. Without gravity, none of these tricks are possible. Instead of balancing, the gyroscope floats. When touched by a string, it moves away.

Commander Bobko also added a piece of chewing gum to unbalance his gyroscope. The lopsided gyroscope wobbled, twisted, and gradually came to a stop as it tried to spin around its new center of mass.

There is one earth trick that did work in space. Commander Bobko attached a string to the gyroscope's cage. With another string, he started the gyroscope spinning. He then rocked the gyroscope back and forth. While swinging, the gyroscope kept its spin axis. He then swung the spinning gyroscope around in circles. The gyroscope's axis reoriented to be perpendicular to the swinging string. This is the most stable configuration as the gyroscope reacts to the inward centripetal force. Don Mc-Monagle repeated this demonstration successfully with the gravitron.

As a final maneuver, Commander Bobko released the gyroscope and began swimming around it. The gyroscope maintained its spin axis regardless of the orientation of its owner. This demonstration was most appropriate. Commander Bobko relied on other delicately balanced almost frictionless gyroscopes to pilot the shuttle. In the last minute before liftoff, the computer-connected gyroscopes of the shuttle's Inertial Measurement Unit were set as the computer noted the shuttle's exact position and orientation. Throughout the flight, shuttle computers consulted these gyroscopes to determine the tilt of the spacecraft relative to its launch position. In the weightlessness of space, the crew and the computers cannot determine the tilt or attitude of the shuttle. There is no downward force to indicate the shuttle's orientation relative to the earth below. If the gyroscopes in the Inertial Measurement Unit were to fail, the crew would have to rely on the positions of stars for attitude calculations.

SPACE RESULTS: GRAVITRON

The performance of the gyroscope on the STS 51D mission gave only a hint of the exciting things that the gravitron would do on STS 54. First, Don McMonagle pushed two gravitrons into each other. At each collision, the spinning gravitron would change direction but keep its orientation, while the nonspinning gravitron would tumble. When two spinning gravitrons collided, both bounced off each other while keeping their axes of spin.

The most exciting gravitron moments came when Don McMonagle attached two and then three gravitrons to a plastic baseball. Holes had been drilled in the baseball so that the gravitrons could be mounted on the same axis and at right angles to each other. Astronaut McMonagle placed one gravitron in each end of the ball. He started each gravitron spinning so that they would be spinning in the same direction and then released the ball. Gradually the spinning disks inside the gravitrons began to turn the gravitron casing, which in turn began to turn the ball. The ball began to spin faster and faster, and then to wobble. Soon the spinning gravitrons were circling the spinning ball. Finally, the circling speed got so great that one gravitron popped off the ball and flew across the cabin. The other gravitron and ball sailed off in the other direction. After this happened, Don McMonagle called the rest of the crew to come to the middeck to see the coupled gravitrons in action. He repeated the experiment several times with the same explosive result.

Then he started the gravitrons spinning so that they were spinning in opposite directions. This time the ball began to turn very slowly. One gravitron pulled free and drifted slowly across the cabin. This time there was no violence. He repeated the experiment with the gravitrons pushed firmly into the holes in the ball. He was also careful to give each gravitron the same amount of spin. As he released the ball, he gave it a push and the ball tumbled with the gravitrons attached. When the gravitrons were spinning in opposite directions, their angular momenta canceled and the ball behaved as if neither gravitron were spinning at all.

Finally, Don McMonagle attached three gravitrons to the ball at right angles. He showed that the ball and gravitrons would tumble if the grav-

itrons were not spinning. Then he began the spinning motion. Each grav-itron was given the same amount of spin and the ball was released. The ball began to turn as the spinning of the gravitrons was transferred to the ball. The ball's stable spin axis was at a 45-degree angle to the axis of each of the spinning gravitrons. The angular momenta of the three gravitrons added together to produce a stable spinning axis directly between the three gravitrons. One gravitron separated from the ball and slowly floated away. The remaining two spinning gravitrons caused the ball to begin spinning around an axis between them. As the spinning increased, one gravitron flew off and the third gravitron flew away with the ball at-tached. The whole demonstration showed dramatically how angular mo-mentum is conserved and how the angular momenta of several gravitrons must be added together when the gravitrons are coupled to form a system.

Commander Gerald Carr of the third Skylab mission first demon-strated the conservation of angular momentum with a toy gyroscope. On January 9, 1974, he started his gyroscope spinning and applied a force to each end of its spin axis using straws. In response to this coupled force, the gyroscope's axis began to turn or precess at an angle perpendicular to the force. He also showed that a slower spinning gyroscope required less force to initiate this precession. Once the straws were removed, the spinning gyroscope ceased to precess. Commander Carr then allowed the two straws to rub against the turning axle. The resulting friction also caused the gyroscope to precess.

SPACE RESULTS: YO-YO

Dave Griggs' yo-yo also depended on its spin to work in space. Before liftoff, physicists had discussed and debated the yo-yo's space behaviors. No one knew exactly how the yo-yo would perform. Some thought that gravity would be needed to keep the string from knotting. Others expected the dynamics of the yo-yo and the string to operate independently of gravity's pull. Still others felt that speed would be critical for successful space yo-yoing. In high-speed practice sessions, several yo-yo strings had broken, so Commander Bobko decreed that there would be no yo-yoing on the flight deck.

Like the top and gyroscope, the yo-yo is a tiny flywheel, storing energy in its spinning motion. It turns this spinning energy into kinetic energy as it moves along the string. On earth its speed is also affected by the direction of the yo-yo motion. A downward-moving yo-yo can go faster because it can exchange its gravitational potential energy for kinetic energy. In space, Dave Griggs yo-yoed in all directions with equal ease. The upward motion, nicknamed the "oyoy," looked just as effortless as the downward yo-yo.

As it turned out, space yo-yoing is elegance in action. Dave Griggs could release the yo-yo at speeds as slow as a few inches or centimeters per second, and the yo-yo would move gracefully along its string. Subtle throwing twists resulted in slight yo-yo motions. The yo-yo automatically returned to its owner. When the yo-yo reached the end of the string, it pushed against the loop. The loop reacted by pushing back on the yo-yo and starting it up the string. When Dave Griggs released the yo-yo's string on its return trip toward his hand, the yo-yo continued winding up the string and the wound-up yo-yo bumped its smiling owner on the nose. When he released the yo-yo on its outward trip, it rolled past its loop and wound up its string while going forward. The space yo-yo must, of course, be thrown and not dropped. There is no gravity force to pull the yo-yo down.

Without a downward force, the yo-yo also refused to sleep. "Sleeping" involves letting a rapidly spinning yo-yo move down the string. The downward gravity force balances a gentle upward rebound as the yo-yo hits the loop at the string's end. After a few moments of sleeping, a jerk on the string will send the yo-yo back upward. In space there is no force to keep the yo-yo at the bottom of the string.

While sleeping tricks, like "walking the dog" or "rocking the baby" are impossible, other tricks become much easier in space. "Shooting the moon" was much easier in space. Dave Griggs noted that his performance was the first time this trick had been done while traveling at 25,000 feet per second (8 kilometers per second). To accomplish this trick, he had to yo-yo horizontally and then vertically without stopping to catch the yo-yo between throws. Dave Griggs readily mastered another trick, the "flying saucer." In this feat, he had to throw the yo-yo sidearm to produce a horizontal spin. The complete trick also involved sleeping the sideways spinning yo-yo—an impossible task in zero gravity.

"Around the world" became the trick that almost worked. It involves swinging the sleeping yo-yo around in circles before returning it to your hand. Unfortunately the middeck was too small for such a performance. If space had been available (or perhaps if the string had been shorter), Dave Griggs might have been able to accomplish this sleeping trick. The inertial forces involved in circular motion would keep the yo-yo pushing outward as it reached the end of the spinning string. Dave Griggs did manage a looping-type yo-yo behavior in space. As the yo-yo returned to his hand, he allowed it to loop around and then sent it back out for another yo-yo journey.

The slowness of the space yo-yo allowed Dave Griggs to observe the dynamics of the yo-yo's motion and to perfect tricks with great care. Only fear of tangling the yo-yo on the shaft restrained his yo-yo performance. One yo-yo tangle did occur during the mission. Dave Griggs released his yo-yo as Jeff Hoffman slung out his slinky. Both got thoroughly twisted together. As Jeff Hoffman commented, "Some toys just don't mix in space."

11

Space collisions

THE MOST ELEGANT and meaningful of all demonstrations is a well-planned collision in microgravity. You can observe momentum and direction of objects before the crash and predict what will happen after the collision. Without a gravity force pulling each object down, things work just as Newton's laws predict they should.

THE BALL AND CUP

The ball and cup is a traditional toy found all around the world. The cup should be just slightly larger than the ball and held by a handle. The ball is attached to the handle with a string.

You will need

- Scissors
- Ruler
- Plastic spoon or popsicle stick
- 2 pieces of kite string about 2 feet in length
- Masking tape

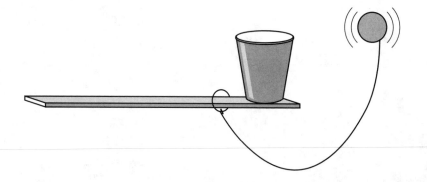

- Ping-pong ball
- Paper bathroom cup

1. Place the ping-pong ball inside the cup. The cup should be just slightly larger than the ping-pong ball. If the cup is taller than the ball, trim it down until the ball just fits in the cup. The smaller the cup, the more challenging your toy will be.
2. Tape the spoon (or popsicle stick) to the bottom of the cup with the bowl curved upward. The spoon handle is the handle of your new toy.
3. Tape one end of the kite string to the ping-pong ball and tie the other to the handle of the spoon at a point near the cup. Wrap the string around the handle until the ball and cup are about 1 foot (30 centimeters) apart.
4. Hold the handle with one hand and let the ball hang down. With a swinging motion, cause the ball to swing into the cup. It must stay in the cup and not bounce out.
5. If you are having difficulty, wrap up more of the string. If it is too easy for you, make the string longer.

Once you can catch the ball, think about how you are using the toy. Can the astronauts use your technique in space? What would happen if they did? How can they catch the ball in the cup? Will this toy be easier or harder to use in space. Try different techniques in playing with the ball and cup. Place the ball in the cup. Push the cup upward and then pull it downward. Watch what happens to the ball. Then push the cup back up to catch the falling ball. Is this a technique that could be used in space?

POOL BALLS AND RACQUETBALLS

The STS 54 astronauts took two pool balls and two racquetballs into space. Both are the same size, but the pool balls have four times the mass of the racquetballs. Susan Helms drew the assignment of pushing the balls into each other and observing what happens after each collision. Since balls roll well, you can experiment with this activity on earth. All you need are four balls of the same size, but different mass. You might try real baseballs and plastic baseballs or small superballs and ping-pong balls.

Roll one of your lightweight balls into the other. Watch what happens after the collision. Then roll one of the light balls into one of the heavy balls. Watch what happens after this collision. Predict what will happen when you roll one heavy ball into the other and then into a stationary light ball.

Place a light ball so that it is touching a heavy ball. Roll the other heavy ball into the first heavy ball. Watch what happens to all three balls. Predict what will happen when you roll the second light ball into the stationary heavy ball next to the other light ball, then try it. Experiment with other collisions using the four balls.

To explain what is happening, think about momentum conservation. The momentum of the first ball you roll must be conserved by the collision. The momentum of the moving ball before the collision equals the momentum of the two moving balls after the collision. When one ball hits another of the same mass, the first ball stops. The second ball leaves the collision with the same velocity and momentum that the first ball had. This is called an *elastic collision* because the balls do not stick together and neither deforms to absorb some of the collision's energy. When the heavier ball hits the lighter one, we see the heavier ball's extra momentum dominating the collision. Also when the lighter ball hits the heavier ball, it rebounds.

KLACKER BALLS

This is a new and surprising space toy for the STS 54 mission. Mario Runco chose the klackers and became quite good at operating them before leaving for space. There are two ways of klacking and each is based on Newton's third law or the conservation of momentum. Try both ways and notice the differences in how the klackers behave. Try to decide which way will be harder to do in space.

Experiment 1: Action/reaction klackers

Hold the two balls horizontally on either side of the handle. Drop the balls at the same time. As they hit, move the handle downward. When they hit on top, move the handle upward. Repeat. Do the klacker balls remain on the same side of the handle or do they change sides? How fast can you do the klackers? Count the number of klacks you can make in 10 seconds.

Experiment 2: Momentum klackers

Hold one ball above the handle and let the other one hang below. Release the top ball. As it swings down, it will hit the lower ball. What happens to the lower ball and to the ball that hit it? With a small turn of the paddle, you can get the moving ball to circle the handle and hit the other ball. What happens then? How can you make this motion continue? How fast can you spin the klackers? Count the number of klacks you can make in 10 seconds? Is this more or fewer klacks than with the action/reaction klackers? Explain why there is a difference. Try both the action/reaction klacker and the momentum klacker activities holding the handle in a vertical position and letting the balls move from side to side. Can either activity be done sideways?

If you have a piece of clay, you can experiment with how the mass of the klackers affects what they do. First, add clay in a ring around one of the klackers. Add enough clay to weigh about as much as one of the balls, then try the momentum klackers activity. What happens this time when the balls hit? Can you make the balls move around like they did in the last activity? Why or why not? Now spread the balls out on either side of the handle. Drop both balls at the same time. Where do the balls meet? After they meet, which ball rises up farther? Can you make the balls klack on the top and bottom like they did when they had the same mass? What is the effect of giving one ball more mass?

SPACE RESULTS: BALL AND CUP

The ball and cup toy was a finalist in the first Toys in Space project. John Casper agreed to give the toy a try on the STS 54 mission. Unfortunately he discovered that this simple earth toy is impossible to master in microgravity. First he tried to swing the ball upward and pull it into the cup. When the ball finally went into the cup, it promptly bounced back out. Then he tried to ease the ball slowly into the cup. But the slightest nudge on the cup caused the ball float back out. Finally, he tried to throw the ball

into the cup. He stretched the ball out in front of the cup and pushed the ball toward the cup. The whole system twisted around. On earth, gravity causes the ball to fall into the cup. In space all John Casper had was the inertia principle. He started the ball moving in a straight line and put the cup in front of it. But when the ball reached the cup, no force held it in. On earth, gravity does more than pulling the ball into the cup. It also keeps the ball in the cup. In space, when the ball hits the cup, it always rebounds away. To make this a workable toy in microgravity, Casper put a piece of tape inside the cup to hold onto the ball.

SPACE RESULTS:
POOL BALLS AND RACQUETBALLS

According to Newton's third law of motion, momentum (mass times velocity) must be conserved. In space Susan Helms found that space collisions between pool balls and racquetballs conserved momentum in all directions. First she pushed the massive pool ball into the light racquetball. After the collision, both balls moved away in the same general direction with the racquetball receiving part of the pool ball's momentum and the pool ball moving at a slower speed. She tried this several times. Each time the collision was slightly off center and each time the two balls also moved apart at an angle. Once again, however, momentum was conserved. If the racquetball moved to the right after the collision, the pool ball would carry an equal amount of momentum to the left.

Next Susan Helms pushed the light racquetball into the more massive pool ball. After the collision, the racquetball bounced backward at a slower speed while the pool ball rolled forward. In this case momentum is conserved, but you must remember that if one ball is rolling forward and the other is rolling backward, you must subtract their momenta to get the total momentum for the system.

These two-ball collisions must conserve energy as well as momentum. The energy of one ball rolling into the other must be shared by both balls after the collision. The more massive ball also has more energy than the lighter ball moving at the same speed. In general, the conservation of momentum determines the directions of the balls after each collision. The conservation of energy determines how fast the balls will be traveling.

Susan Helms discovered that it was impossible to perform the three ball tricks because she could not keep two balls together long enough to hit them with the third ball. To provide a modified demonstration, she taped a pool ball and a racquetball together. Then she hit the pair of balls with either a racquetball or a pool ball. In effect, she reduced the demonstration to a version of the two-ball collision. This time the stationary object was more massive than either of the colliding balls. When the racquetball hit the other two balls, the result was much the same as when the racquetball hit the pool ball. When the pool ball hit the other two balls, it was almost like the pool ball hitting another pool ball (since the racquetball is much lighter). The colliding pool ball virtually stopped and the other pool ball with the racquetball attached moved away with the momentum and kinetic energy of the colliding pool ball.

SPACE RESULTS: KLACKER BALLS

Klacker balls illustrate momentum conservation more dramatically than the pool balls and racquetballs do. Since the two klacker balls have the same mass, it is easy to see this conservation in action. It is harder to predict which klacker behaviors would be more challenging in microgravity.

Mario Runco began by doing the action/reaction klacker demonstration. He moved his hands up and down rapidly as the klacker balls collided above and then below the handle. This procedure worked very well. After generating a very rapid klacker motion, he released the handle. The klackers moved off in the direction that the balls were moving at the time he let go. The balls moved off in a straight line instead of klacking together again.

The second klacker motion proved impossible to sustain. On earth it is easy to release a klacker ball from above and watch as it hits the bottom ball, giving its momentum to the second ball. The second ball then circles around the handle, leaving the first ball hanging down. With a circular motion of the wrist, you can keep this motion going easily on earth. In space, there is no force to keep the bottom ball down. The small wrist motions used to send the top ball around also cause the bottom ball to rise. After several minutes of earnest effort, Mario Runco was unable to sustain this klacking motion. He could get about two successful hits before the two balls would begin spinning around the handle without contact. He thought that with enough time, he might have learned to master this motion in microgravity, but it wasn't easy. Gravity definitely plays a part in stabilizing the klackers when they are circling in the same direction.

12

Space games

ADAPTING EARTH GAMES to freefall conditions is perhaps the most challenging of all space activities. A game is more than a toy or demonstration. A game has rules. In a game you do certain things to win whether or not you are playing with an opponent. Thus far, space demonstrations have been limited to games for one player. The middeck is too crowded for two player and team sports. These must wait for a more roomy space station. All of the games chosen for space work well on earth. The trick is to redesign the action and rewrite the rules for microgravity.

Crucial momentum effects are often unnoticed when a game is played in a world with weight. Gravity's downward pull and the resulting friction force on moving objects can mask the momentum of a moving toy. In space, inertia rules. Objects in motion do stay in motion regardless of astronaut wishes. Momentum is conserved and cannot be transferred to the massive earth. Astronauts mastering a space game quickly discover that the slow motion of a big astronaut can become the fast getaway of a tiny ball, horseshoe, or jack.

TARGET PRACTICE

This is a game you can make after a trip to the fabric store. The target is a circle 8 inches (20 cm) wide covered in pile Velcro. A bull's eye is drawn on the fabric. The balls are the size of racquetballs, but much lighter. Each ball is wrapped in four strips of hook Velcro. About 90 percent of the time, the Velcro ball will stick to the target on impact. Smaller versions of this toy are available in toy stores. Once you have constructed your ball and target, try these activities.

Experiment 1: Hitting the target

Place the target on the wall. Stand about 6 feet (about 2 meters) away. Throw four balls underhanded. Record your score. Then throw four balls overhanded. Record your score. Which throwing style is more accurate? What path did your ball take as you completed each throw? Which throwing style might work better in space?

Experiment 2: The suspended target

Hang the target from a string in the center of the room. Throw the ball toward the target. What happens to the target when you hit the center? What happens to the target when you hit its edge? What happens when you hit the target with a faster ball? What do you think will happen to a floating target in space?

Experiment 3: The curve ball

Try to put a spin on the ball as you throw it. For a curve ball, the spin is to the side. (A fast ball has a bottom spin and a sinker has a top spin). See if you can put enough spin on the ball to change the direction of its motion.

The curve ball shows the Magnus effect at work. When the ball spins, a pressure difference occurs between the leading and trailing sides of the ball. On the leading side (which is the right side in a curve ball thrown by a right-hander), the surface of the ball moves faster through the air than on the trailing side. Because the drag force is greater at larger velocities, there is more force on the leading side than on the trailing side. Since the leading side is on the right and the trailing side in on the left, this drag force causes the ball to curve to the left. Do you think a curve ball will curve in microgravity?

HORSESHOES

The horseshoes taken into space are the heavy plastic variety—bright green and pink in color. The post is yellow plastic and has a circular bottom. The other side of the horseshoe post was used as the dart board in the target practice game. You can use any horseshoe set in trying these earth challenges. The heavier the horseshoes, the easier they are to throw.

Experiment 1: Ringing the post

Place a plastic post on the floor inside (or in the ground outside). Stand about 6 feet (2 meters) from the post and toss the horseshoes toward it. You toss one and then your partner tosses one. In scoring, you get five points for each ringer (horseshoe around the post), three points for each horseshoe touching the post, and one point for the horseshoe closest to the post. Notice the path that your horseshoes take toward the post and your most successful throwing strategy.

Experiment 2: The horizontal post

Hang the post from a wall or use a post already sticking out of a wall. Toss the horseshoes so they hang on the post. Describe the best strategy for making a ringer.

Experiment 3: Spinning horseshoes

Place the post on the floor inside (or in the ground outside). Stand about 6 feet (about 2 meters) from the post. Try spinning the horseshoe like a Frisbee. How does it fly? How successful are you in making a ringer?

Experiment 4: Heavy horseshoes

Tape two horseshoes together. Compare tossing them with tossing a single horseshoe. Observe if it is easier to make a ringer, to bounce off the post, or to lean against the post.

After trying these activities, predict how each will work in microgravity. Do you think that Susan Helms will be able to make a ringer in space?

BALL AND JACKS

To play jacks you first spread the jacks out on a table. You then bounce the ball, pick up a jack, and catch the ball with the same hand. Transfer the jack to your other hand. Repeat this procedure until you pick up all the jacks. You have then finished your "onezes". Next you repeat the procedure and pick up two jacks at a time. Continue this process until you pick all of the jacks up at once. Once you miss the ball, move a jack you are not picking up, or drop a jack, you must give up your turn.

To adapt jacks for weightlessness, you must decide how gravity affects the game.

- Can Rhea Seddon bounce the ball in space?
- How can she get the bounced ball to return to the table?
- How fast should she throw the ball in space?
- How should she release the jacks in space?
- Will she need both hands in controlling the jacks?
- When should she stop playing and collect runaway jacks?

Gravitational potential energy depends on the distance a toy can fall and on the force that is pulling the toy downward. In the reference frame of an orbiting shuttle, there is no effective downward force and, therefore, no gravitational potential energy. For this reason, the rules in a game of space jacks must be rewritten. In space the ball does not slow down as it travels upward, because it does not convert its kinetic energy to gravitational potential energy. It must bounce off the ceiling or wall to change its direction and return. The ball's speed and kinetic energy are constant. Space jack expert, Rhea Seddon, had to adjust her reflexes to catch a ball moving at a constant speed rather than slowing down as it rose. A jack bumped by mistake would also suddenly acquire kinetic energy and drift away. On the positive side, Rhea Seddon could throw the ball so slowly that she could collect all the jacks before the ball bounced off the wall and returned.

BASKETBALL

No sport is more dependent on gravity than basketball. Gravity pulls the basketball through the hoop and limits the hang time of the sport's great jumpers. Gravity determines how shots are made and holds players to the floor. Every aspect of basketball is ruled by gravity's downward force. Try these activities as you rewrite this famous sport for the freedom of freefall. The flight model has a foam ball and a plastic hoop held to the wall with suction cups.

Experiment 1: Making a basket

Using the suction cups, secure the basket to a wall. Practice throwing the ball until you make a basket. Keep your distance the same (like making free-throws). Ask your partner to draw the path of the tossed ball as it enters the basket. You can attach the hoop to a chalkboard and draw each path. What causes the path to curve? Where do you aim when you throw the ball?

Experiment 2: Bouncing baskets

In this game, you must bounce the ball through the basket. You can bounce the ball off the floor, ceiling, or any wall. You must remain about 6 feet (about 2 meters) from the basket when you shoot. Notice the ball's path for a successful shot off the floor, ceiling, and wall above the basket. Which paths are most influenced by gravity? Which technique might work in space?

Experiment 3: Basket heights

With suction cups, attach the basket to the wall at each of these heights and make a basket: 20 inches (50 cm) off the floor, 40 inches (about 1 meter) off the floor, 60 inches (about 1.5 meters) off the floor, and 80 inches (about 2 meters) off the floor. You must remain 80 inches (about 2 meters) from the wall when you throw. What is the easiest height to make a basket? the hardest?

Experiment 4: Writing the rules

Think about how each basketball behavior will be different in space:

- Dribbling the ball
- Walking with the ball
- Guarding an opponent
- Dunking the ball
- Goaltending
- Standing in the lane too long

Now make rules for a playable half-court basketball game in space with two players on each team.

SPACE RESULTS: TARGET PRACTICE

Susan Helms discovered that target practice is much easier in microgravity. On earth, you must throw the ball fast enough to reach the target. Otherwise gravity will pull the ball to the ground. In space the ball travels in a straight line and can be thrown at any speed. To score a bull's eye, Susan Helms just aimed directly at the target.

To increase the challenge of space darts, she tried to spin the ball just to see if it would curve. After several tries, the ball did seem to curve before reaching the target. Balls are certainly better than darts for space target practice. Arrow-shaped darts are more dangerous in a free-floating environment. Balls are safe to throw and can be tossed as sinkers and curve balls to make the game more interesting.

A floating target is definitely challenging. When the ball hit the bull's eye, the floating target drifted away. When the ball landed near the target's edge, the target would flip over and over. The ball hitting in the center brought its momentum to the stationary target. After it stuck to the target, the ball and target shared this momentum and moved away. A ball hitting on the target's side brought angular momentum to the target. To conserve angular momentum, the target had to spin.

As a final demonstration, Susan Helms hit the floating target with one ball near the side and started the target flipping. Then she hit the target on the other side of the bull's eye with the second ball and the target stopped spinning.

SPACE RESULTS: HORSESHOES

Horseshoes must be thrown knowing that linear momentum and angular momentum are conserved upon impact with the post. When the post is mounted on a locker, the challenge is to throw the horseshoe so that the linear momentum is minimal at the point of impact. Otherwise the horseshoe will bounce off of the post. Any spinning of the horseshoe must also be carefully planned. A spinning horseshoe can spin off the post as it conserves angular momentum.

Susan Helms tried several practice shots to find the best way to toss the horseshoes. If the horseshoe hit the post or base, the horseshoe would bounce off. Finally she managed to throw a horseshoe so that the small hook of the horseshoe tip caught on the post. The horseshoe began circling the post and for almost five minutes the circling continued. Gradually the horseshoe moved upward on the post and slowed down. Finally it left the post while still spinning. This was Susan Helm's only ringer. Every other horseshoe she threw bounced off the target.

With a floating post, the horseshoes hitting the post caused the post to tumble. As an experiment, Susan Helms threw the horseshoes to hit the post at three different spots: close to the bottom of the post, halfway up the post, and near the tip of the post. When the horseshoe hit the bottom of the post, it caused the post to drift away with a slight tumbling motion. When the horseshoe hit halfway up the post, the tumbling motion increased. When the horseshoe hit at the top of the post, the post tumbled wildly and the horseshoe flew away. In each case conservation of angular momentum caused the result. The farther the horseshoe hit outward on the post, the more angular momentum it gave to the post and the greater the tumbling motion.

SPACE RESULTS: BALL AND JACKS

Imagine opening your hand while holding 11 jacks. Jacks bump into one another in your tightly closed fist. Each jack sticks to your unfolding fingers. Your hand moves slightly as you spread your fingers. The result, jacks fly apart like fire flies. No gravity force confines the wayward jacks to a tabletop.

Rhea Seddon first mastered the technique of very gentle jack placement, always aware that a small nudge and a tiny bit of momentum transfer would send the jacks flying. It would have been better to hold the jacks cupped in both hands and to open the hands away from the stationary jacks. Unfortunately two hands holding the jacks leaves no hands to toss the ball. Whenever she played with jacks on camera, two other astronauts watched the tiny jacks to prevent them from drifting into the middeck equipment.

Rhea Seddon also tackled the ball tossing problem. This proved to be an interesting question for many young scientists. Most agreed that a tossed ball would not fall. Yet many children also believed that a ball would not bounce in space. For these students, the space jacks demonstration had many surprises.

Rhea Seddon decided to put a bounce and a minimum throwing speed into her space jacks rules. It was possible to tap the ball so lightly that the jacks could be rearranged many times before the ball bounced off the wall and returned. A minimum ball speed was necessary for space jacks to be as challenging as its terrestrial equivalent. Ball catching involved a slight space adjustment. On earth a tossed ball's speed is constantly changing. In space it is constantly the same. A space jacks player must keep watch on the bouncing ball and not let terrestrial catching experiences cause a premature grab.

Floor and ceiling bounces are possible and equally easy according to Rhea Seddon. The closer the surface is, of course, the more challenging the game, because the ball returns more quickly (assuming a constant tossing speed). For astronauts with feet in toeholds on the floor, the ceiling is the nearest bouncing surface. The ceiling is also better for making movies because the ball remains in view for most of the jacks game. By the last day of the mission, Rhea Seddon reached her "elevenzes" as she collected all the jacks before grabbing the returning ball.

In space, jacks made excellent momentum models. When the ball hit a much lighter jack, the jack flew off at many times the ball's speed. Because momentum increases with more mass or speed, conservation resulted in the lighter jack acquiring a greater speed from the more massive, slower moving ball.

A jack would also spin like a tiny top in space. It drifted when touched by a finger, but always kept its spin axis orientation and never tumbled. As Jeff Hoffman commented, "Whether a satellite or a tiny jack, a gyroscope is a gyroscope."

SPACE RESULTS: BASKETBALL

Earth's gravity determines the parabolic arc of a basketball. With a little practice you can predict the shape of this arc and make a basket on Earth. In space, the basketball travels in a straight line (Newton's first law). Greg "Space" Harbaugh focused on three basketball challenges: the banked shot, the free throw, and the slam dunk.

Earth's easy banked shot is impossible in microgravity. Greg Harbaugh tried over a dozen shots at close range and could never get the basketball to bounce off the backboard and into the basket. A normal banked shot hits the backboard and comes back to you. Even when he was high over the basket, his banked shot touched the basket, but did not have enough force to go through the netting. The problem is more apparent when you draw the angles involved. When the ball hits the wall, it bounces off at an angle

equal to the incoming angle. Greg Harbaugh must be well above the backboard to throw a ball that will bounce downward through the hoop after hitting the board. And when he did get the angle correct, the basketball bounced off the wall and hit the hoop and net at an angle. Instead of staying in the net, the ball bounced off at an angle. It became obvious that the ball must travel straight into the hoop or it will bounce off. Greg Harbaugh used this lesson as he attempted the free throw.

Greg Harbaugh tried his free throw from a distance of about 3 feet (about 1 meter). Even at this distance, the banked approach would not work and the ball would never enter the basket without being redirected. So he bounced the basketball off the ceiling. After about 15 on-camera practices, the ball actually hit the ceiling and went straight into the basket. The most surprising result was the speed required for the basketball. The ball must be moving quickly or it will drift off the top of the hoop without dropping in. After each basket, the ball was trapped in the net. Thus the lack of a downward pull leaves space basketball without a banked shot or a swish. Even the free throw requires a surface above the basket to redirect the ball's motion.

The slam dunk is another matter. This is where Space Harbaugh earned his name over and over again. On earth dunking requires height and jumping strength. In space anyone can float over and push the ball through the basket. The challenge is to dunk the ball in style. Greg Harbaugh created two techniques: the "360 × 3 in the tuck position" and the "360 × 4, 5, or 6 in the layout position." In the second delivery, he pushed off from the far wall of the middeck with his hands and body in as straight a position as possible. He then turned in circles as he floated toward the basket. Four, five, and even six spins were possible. At the last minute, he raised his hand holding the basketball and pushed the ball through the hoop. His greatest challenge was to avoid hitting the wall with a loose arm or leg or reaching the basket facing the wrong direction.

The multiple 360 in the tuck position was a much more challenging delivery. Space Harbaugh began in the corner of the middeck with a push off the wall. After the push, he pulled his arms and legs into a ball and turned around and around as he floated toward the hoop. In this position, bouncing off a wall became necessary as a midcourse correction on the way to the goal. Finally he rolled close enough to reach out a hand and push the ball through the hoop.

The STS 54 crew never attempted a team basketball game. In a middeck with less than 8 feet of moving room in any direction, a game with more than one player would degenerate to a chaos of floating arms and

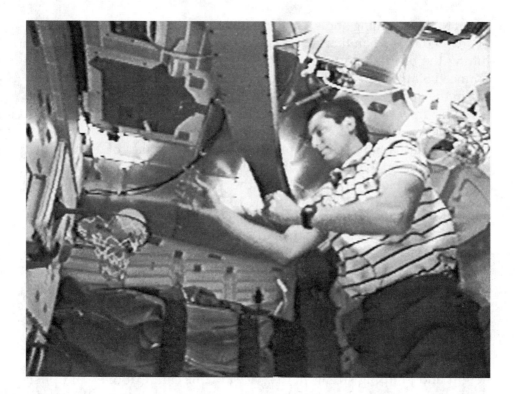

legs. With no traction to start or stop motions, basketball would become an accidental contact sport. Players would need a gymnasium the size of a large classroom to have freedom to move and maneuver. Speed and jumping strength would be replaced by coordination and timing as the skills of a space basketball superstar.

13

Coming home

ALL ELEVEN TOYS completed their final space performances on the last full day of the STS 51D mission. Twenty-four toys were captured on video during the STS 54 flight. These demonstrations began on day 3 and extended to day 5 of the mission. After filming or taping, each toy was carefully stowed in the toy carrying bag inside a storage locker on the shuttle's middeck.

Seats made useless by weightlessness were once again assembled and attached for reentry. Astronauts donned special gravity pants. Once inflated, the pants pushed firmly against astronaut legs and forced blood to flow back toward the head and heart. Without these pants, an astronaut might black out on reentry.

To leave orbit, the commander turned the shuttle around and fired its small maneuvering engines to reduce the craft's speed. A loss of only 300 feet per second (92 meters per second) in speed started the descent.

As reentry began, the commander turned the shuttle back to a forward position. He tilted the heavily heat-shielded underside to take the brunt of reentry. The shuttle dropped 125 miles (200 kilometers) in 30 minutes. At the 75 mile (120 kilometer) altitude, 30 minutes before landing, the shuttle entered earth's upper atmosphere. During the terrific heat build up of the next 12 minutes, underbody temperatures reached 2,500° F (1370° C), while wingtip temperatures rose to 3000°F. (1650°C). The cabin remained cool because of the remarkable thermal insulating tiles that cover the shuttle's exterior.

During the last stages of reentry, the shuttle's plasma wake prevented ground communications. Five computers—each checking on the others—controlled the speed and angle of descent.

The shuttle lands like a glider. It has no engines that can fire for a second pass. Once reentry begins, the shuttle must land somewhere. The commander uses controls similar to those in an aircraft cockpit to guide the craft. The shuttle computers measure wind speeds and descent angles and make the necessary adjustments in the shuttle's wing and tail flaps. This combination of astronaut and computer control is called "flying by wire."

The STS 51D and STS 54 flights both landed in Florida instead of in California. No motors sounded as the shuttle dropped toward the runway at the Kennedy Space Center. The descent angle was seven times as steep as a commercial airliner approach. But the landing speed of 200 miles per hour (320 kilometers per hour) was only a little faster than a jet's landing velocity.

The rear tires touched down first followed by a lowering and touchdown of the front tire. The STS-51D mission was the first to land at the Kennedy Space Center in a significant 8-knot crosswind. To stabilize the shuttle for landing, Commander Bobko applied pressure to the right tire. The tire's brakes overheated and locked. The tire's skidding further increased the temperature and led to a blow out just as the shuttle *Discovery* rolled to a stop. This final mishap provided one last science demonstration for this eventful mission. With a parachute in tow, the shuttle *Endeavour* made a perfect landing for the STS 54 mission.

The total mission elapsed time for STS 51D was 6 days, 23 hours, and 56 minutes. Commander Bobko had guided his spacecraft through over 100 earth orbits. The STS 54 mission lasted 5 days, 23 hours, and 38 minutes and completed 96 earth orbits.

Once unpacked, the space toys left Florida for teaching roles in a display at the Houston Museum of Natural Science. Senator Garn's paper airplane journeyed to the Smithsonian's Air and Space Museum in Washington, D.C. The toys that rode on the shuttle *Endeavour*, but were never demonstrated, were sent to the Zero Gravity Testing Facility at the Johnson Space Center. These toys completed their mission in brief 20-second demonstrations aboard the KC-135.

After the STS 51D mission, many new toys were proposed and new activities were suggested for the old toys. Most of these requests from teachers and students became the activities of the STS 54 mission. Multiple gyroscopes, two-tone magnetic marbles of different sizes, balloon helicopters, and boomerangs were all suggested after the first mission.

The list of toys that can teach science in space is as boundless as the creativity of human toymakers and the eagerness of children to investigate new toy behaviors. Such childhood curiosity is a resource often untapped in science classes.

A reporter and camera crew visited a typical fourth-grade classroom to tape children discussing the behaviors of the toys in space. The adults watched as the youngsters predicted toy behaviors in space and then observed the videotaped space toy motions. Without fear or hesitation, they considered inertia, momentum, centripetal force, and other science concepts in explaining the behaviors of favorite toys. The room filled with "why's" and "what if's" as children suggested more space demonstrations, experimented with new toy behaviors, and recommended new space toys for future missions.

Appendix A

The toy operations crews

STS 51D CREW

Commander Karol Bobko (Colonel, USAF) was a member of the first graduating class of the U.S. Air Force Academy and became its first graduate to fly in space. He has received two NASA Exceptional Service Medals and six JSC Group Achievement Awards. He was also presented the Air Force Distinguished Flying Cross and two Meritorious Service Medals (1970 and 1979) as well as the Air Force Academy Jabara Award

The STS 51D crew: (top row) Don Williams, Karol Bobko, David Griggs, and Senator Jake Garn; (bottom row) Jeff Hoffman, Rhea Seddon, and Charles Walker.

for 1983. Colonel Bobko was pilot for shuttle mission STS 6, launched on April 4, 1983—the maiden voyage of the *Challenger*.

Colonel Bobko commanded the STS 51J flight for the Department of Defense in October 1985. It was the maiden voyage of the *Atlantis*. With the completion of this flight Colonel Bobko logged a total of 386 hours in space.

Karol Bobko was born in New York City in 1937. He and his wife, Dianne, have two children, Michelle and Paul, born in 1963 and 1965. Without Commander Bobko's enthusiasm, the Toys in Space project would never have gotten off the ground. His eagerness to demonstrate the toys and to schedule camera practice sessions were invaluable in the development of a quality film and meaningful results for the first Toys in Space flight.

Dr. Jeffrey A. Hoffman (NASA astronaut) graduated first in his class from Amherst College in 1966 and received his doctorate in astrophysics from Harvard University in 1971. He has designed x-ray astronomy equipment for rocket and satellite experiments and has been in charge of prelaunch design and scientific analysis of x-ray and gamma ray data from the orbiting HEAO-1 satellite. Dr. Hoffman has authored or coauthored more than 20 papers on x-ray bursts since their discovery in 1976. While with NASA, he has worked on shuttle guidance, navigation, and flight control systems.

Following the STS 51D flight, Dr. Hoffman worked as the Astronaut Office Payload Safety Representative and on the development of a high-pressure space suit for use on space station *Freedom*. Dr. Hoffman made his second space flight as a mission specialist on STS 35 in December 1990. This spacelab mission featured the ASTRO-1 ultraviolet astronomy laboratory, a project on which Dr. Hoffman has worked since 1982. Dr. Hoffman made his third space flight as payload commander and mission specialist on STS 46 in July 1992. On this mission, the crew deployed the European Retrievable Carrier, a free-flying science platform, and carried out the first test flight of the Tethered Satellite System. At the time of this publication, Dr. Hoffman is training as a mission specialist on STS 61. This is the mission scheduled to repair and refurbish the Hubble Space Telescope.

Dr. Hoffman was born in Brooklyn, New York, in 1944. He and his wife, Barbara, have two children, Samuel and Orin, born in 1975 and 1979. He enjoys skiing, mountaineering, hiking, bicycling, swimming, sailing, and music. His thoughtful consideration of each toy and insights concerning the physics involved resulted in a series of unique and interesting experiments. His ideas and firsthand observations have added a new dimension to the Toys in Space project. All of the toys chosen by Dr. Hoffman returned to space in the STS 54 mission.

Pilot Donald Williams (Commander, USN) has a degree in mechanical engineering from Purdue University and graduated from the U.S. Naval Test Pilot School at Patuxent River, Maryland, in 1974. As a NASA astronaut, he has been a test pilot for the Shuttle Avionics Integration Laboratory and the Deputy Manager, Operations Integration, for the Space Shuttle program.

From July 1985 through August 1986 he was the Deputy Chief of the Aircraft Operations Division at the Johnson Space Center, and from September 1986 through December 1988 he served as Chief of the Mission Support Branch within the Astronaut Office. On his second space flight he commanded STS 34, which successfully deployed the *Galileo* spacecraft on a journey to explore Jupiter.

Commander Williams was born in Lafayette, Indiana, in 1942. He and his wife, Linda, have two children, Jonathan Edward and Barbara Jane, born in 1974 and 1976. He enjoys all sports activities and his interests include running, woodworking, and photography. His talent as a paddleball operator and juggler produced space toy demonstrations that would not otherwise have been possible. His humorous approach to the toy project, especially his unique position as the first trainer for the flipping mouse (Rat Stuff), heightened the enthusiasm of everyone involved.

David Griggs (NASA astronaut) has been awarded the Navy Distinguished Flying Cross, 15 Air Medals, 3 Navy Commendation Medals, Navy A Unit Commendation, the NASA Achievement Award, and the NASA Sustained Superior Performance Award. He graduated from Annapolis in 1962 and from the U.S. Naval Test Pilot School at Patuxent River, Maryland, in 1967. He became a research pilot at the Johnson Space Center in 1970 and an astronaut in 1979.

David Griggs was born in Portland, Oregon, in 1939. He and his wife, Karen, have two children, Allison Marie and Carre Anne, born in 1971 and 1974. His skill with the yo-yo and eagerness to demonstrate its operation resulted in some of the most entertaining educational footage produced in the Toys in Space project.

David Griggs died on June 17, 1989, near Earle, Arkansas, in the crash of a vintage World War II airplane. At the time of his death, he was in flight crew training as pilot for STS 33, a dedicated Department of Defense mission, scheduled for launch in August 1989.

Dr. Rhea Seddon, M.D., (NASA astronaut) received her bachelor of arts degree from the University of California, Berkeley, in 1970 and her doctorate of Medicine from the University of Tennessee College of Medicine in 1973. She became an astronaut in 1979. During her general surgery residency in Memphis, she pursued her interest in surgical nutrition. She has served as an emergency room physician in Mississippi and Tennessee, and in Houston during her spare time. As an astronaut she has worked with the Shuttle Avionics Integration Laboratory, the Flight Data File, and the Shuttle medical kit. She has also been a rescue helicopter physician.

After the STS 51D mission, Dr. Seddon served on the crew of STS 40 Spacelab Life Sciences (STS 1), a dedicated space and life sciences mission in June 1991. As of this writing, Dr. Seddon is payload commander of STS 58, the Spacelab Life Sciences 2 mission.

Dr. Seddon was born in 1947 in Murfreesboro, Tennessee, and is married to astronaut Robert (Hoot) Gibson. They have a child named Paul, born in 1982. The enjoyment that she showed in demonstrating the jacks and slinky is shared by all who watch the Toys in Space film.

Senator Jake Garn (Payload Specialist and Senator from Utah) was elected to the Senate in 1974. Prior to his election he served as mayor of Salt Lake City and a pilot in the Utah Air National Guard. His congressional responsibilities focus on finance and appropriations committees and subcommittees. Senator Garn completed his shuttle astronaut training in February 1985, just prior to the mission liftoff.

Senator Garn was born in Richfield, Utah, in 1932, is married and has seven children. He agreed to fly a paper airplane for the Smithsonian National Air and Space Museum. His willingness to donate time to this educational project provided a toy that earth-based participants could readily duplicate for their own experiments.

Charles Walker (Payload Specialist) from McDonnell Douglas, operated the Continuous Flow Electrophoresis System located on the shuttle's middeck. He graduated from Purdue University in 1971 with a bachelor of science degree in aeronautical and astronautical engineering. Charles Walker also flew with the electrophoresis equipment on shuttle mission STS 41D. Although he had no toy to demonstrate, Charles Walker experimented enthusiastically with the toy-like behaviors of space bubbles and the on-board Velcro modifications of Rat Stuff's feet.

STS 54 CREW

John H. Casper (Colonel, USAF) was commander of the Shuttle *Endeavour's* third space mission. Selected to be an astronaut in 1984, this was

The STS 54 crew: Mario Runco, John Casper, Don McMonagle, Susan Helms, and Greg Harbaugh.

John Casper's second flight. Colonel Casper also served as pilot on *Atlantis'* STS 36 mission in February 1990. He received a bachelor of science degree in engineering science from the U.S. Air Force Academy in 1966 and a master of science degree in astronautics from Purdue University in 1967. He is a 1986 graduate of the Air Force Air War College.

Colonel Casper and his wife, Chris, have two children, Robert and Stephanie, born in 1982 and 1985. Colonel Casper's commitment to education made the second Toys in Space mission possible. His enthusiasm for sharing the experience with Houston children led to a very successful paper flyer contest prior to the mission. His wife Chris is the assistant principal in a local elementary school and a strong supporter of NASA's education programs.

Donald R. McMonagle (Colonel, USAF) was selected as an astronaut in 1987 and made his first flight as a mission specialist aboard *Discovery* on STS 39 in April 1991. He was pilot for the STS 54 flight. He holds a bachelor of science degree in astronautical engineering from the U.S. Air Force Academy and a master of science in mechanical engineering from California State University, Fresno. He graduated from pilot training at Columbus Air Force Base, Mississippi, in 1975.

Colonel McMonagle's interest in the spinning toys led to the most dramatic demonstrations in all of the Toys in Space experiments. From the come-back can to the coupled gravitron's, Don McMonagle focused on producing demonstrations that would be easy to understand and meaningful to audiences of all ages. His coupled gravitrons flying off the ball is destined to be a classic demonstration of the conservation of angular momentum.

Gregory J. Harbaugh (NASA astronaut) was mission specialist on STS 54. Before being selected as an astronaut in 1978, Harbaugh held engineering and technical management positions in various areas of space shuttle flight operations—particularly data processing systems—and supported real-time shuttle operations from the JSC Mission Control Center for most of the flights from STS 1 to STS 51L. He flew as a mission specialist on STS 39 and was responsible for operation of the remote manipulator system robot arm and the Infrared Background Signature Survey spacecraft.

Greg Harbaugh, who considers Willoughby, Ohio his hometown, received a bachelor of science degree in aeronautical and astronautical engineering from Purdue University in 1978 and a master of science degree in physical science from the University of Houston, Clear Lake, in 1986.

The athletic prowess of "Space" Harbaugh will long be remembered in science classrooms around the nation. His difficulty in making the simplest of basketball shots illustrates to all students how much the gravity force affects their everyday lives. Although less familiar, Greg Harbaugh's experiments with magnetic marbles may be more valuable for students learning about forces. With the two-color marbles, we saw magnetic field effects that we could never see in magnetic marbles rolling on a terrestrial table.

Mario Runco, Jr. (Lieutenant Commander, USN) was a mission specialist on STS 54. He received a bachelor of science degree in meteorology and physical oceanography from City College of New York in 1974 and a master of science degree in meteorology from Rutgers University, New Brunswick, New Jersey, in 1976. After graduating from Rutgers, Runco worked for a year as a research hydrologist conducting groundwater surveys for the U.S. Geological Survey on Long Island, New York. He worked as a New Jersey State Trooper until entering the U.S. Navy in 1978 and being commissioned that same year.

Mario Runco has served in various Navy posts, being designated a Naval Surface Warfare Officer and conducting hydrographic and oceanography surveys of the Java Sea and Indian Ocean before joining NASA. He served as a mission specialist aboard *Atlantis* on STS 44 in November 1991.

Mario Runco's creativity led to a series of very exciting demonstrations. His persistence with the friction car produced enough footage for analysis by students of all ages. His experiments with the klackers are classic—from his determined efforts to make both klacking strategies work to his brilliant demonstration of chaotic motion caused by spinning the klackers around their handle. Children of all ages also appreciate Mario Runco's smiles and obvious fascination with the behavior of each toy.

Susan J. Helms (Captain, USAF) was the third mission specialist on STS 54. From Portland, Oregon, she was selected as an astronaut in 1990. Captain Helms received a bachelor of science degree in aeronautical engineering from the U.S. Air Force Academy in 1980 and a master of science degree in aeronautics and astronautics from Stanford University in 1985. She was an F-16 weapons separation engineer at Eglin Air Force Base, Florida, and served as an assistant professor of aeronautics at the U.S. Air Force Academy. In 1987 she attended Air Force Test Pilot School at Edwards Air Force Base, California, and worked as a flight test engineer and project officer on the CF-18 aircraft at CFB Cold Lake, Alberta, Canada. STS 54 was her first space shuttle flight.

Susan Helms was in charge of taping the toys in space and was devoted to getting the best footage possible. Her attention to camera angles and composition resulted in the excellent footage produced in the STS 54 mission. She also worked to provide the best possible demonstrations for the colliding balls and the balls hitting the target. Her 5-minute ringer in space horseshoes is a classic in unexpected toy behaviors.

Appendix B

Teacher Resource Center Network

TO MAKE ADDITIONAL information available to the education community, NASA's Education Division has created the NASA Teacher Resource Center Network. Teacher Resource Centers (TRCs) offer a wealth of information for educators: publications, slides, audio cassettes, videocassettes, telelecture programs, computer programs, lesson plans, activities, and lists of publications available from government and nongovernment sources. Because each NASA Field Center has its own areas of expertise, no two TRCs are exactly alike. Telephone calls are welcome if you are unable to visit the TRC that serves your geographic area. A list of centers and the geographic regions they serve follows.

Three videocassettes on the Toys in Space program are available: one from the STS 51D mission and two from the STS 54 mission—the *Physics of Toys* lecture from space and a Toys in Space summary. They can be obtained by contacting one of the Teacher Resource Centers listed in this appendix, or the videocassette and toys used throughout this book can be ordered through TEDCO, Inc. Information on ordering through TEDCO appears at the end of this appendix.

If you live in:

Alaska, Arizona, California, Idaho,
Montana, Nevada, Oregon, Utah,
Washington, Wyoming

Write to:

NASA Teacher Resource Center
Mail Stop TO-25
NASA Ames Research Center
Moffett Field, CA 94035
(415) 604-3574

Connecticut, Delaware, District of Columbia, Maine, Maryland, Massachusetts, New Hampshire, New Jersey, New York, Pennsylvania, Rhode Island, Vermont	NASA Teacher Resource Laboratory Mail Code 130.3 NASA Goddard Space Flight Center Greenbelt, MD 20771 (301) 286-8570
Colorado, Kansas, Nebraska, New Mexico, North Dakota, Oklahoma, South Dakota, Texas	NASA Teacher Resource Room Mail Code AP-4 NASA Johnson Space Center Houston, TX 77058 (713) 483-8696
Florida, Georgia, Puerto Rico, Virgin Islands	NASA Educators Resource Laboratory Mail Code ERL NASA Kennedy Space Center Kennedy Space Center, FL 32899 (407) 867-4090
Kentucky, North Carolina, South Carolina, Virginia, West Virginia	Virginia Air and Space Museum NASA Teacher Resource Center 600 Settler's Landing Rd. Hampton, VA 23669 (804) 727-0800
Illinois, Indiana, Michigan, Minnesota, Ohio, Wisconsin	NASA Teacher Resource Center Mail Stop 8-1 NASA Lewis Research Center 21000 Brookpark Rd. Cleveland, OH 44135 (216) 433-2017
Alabama, Arkansas, Iowa, Louisiana, Missouri, Tennessee	NASA Teacher Resource Center Alabama Space and Rocket Center Huntsville, AL 35807 (205) 544-5812
Mississippi	NASA Teacher Resource Center Building 1200 NASA John C. Stennis Space Center Stennis Space Center, MS 39529 (601) 688-3338
The Jet Propulsion Laboratory (JPL) serves inquiries related to space and planetary exploration and other JPL activities.	NASA Teacher Resource Center JPL Educational Outreach Mail Stop CS-530 4800 Oak Grove Dr. Pasadena, CA 91109 (818) 354-6916

California (mainly cities near Dryden Flight Research Facility)	NASA Dryden Flight Research Facility Public Affairs Office (Trl. 42) NASA Teacher Resource Center Edwards, CA 93523 (805) 258-3456
Virginia and Maryland's Eastern Shores	Wallops Flight Facility Education Complex-Visitor Center Building J-17 Wallops Island, VA 23337 (804) 824-1176

To offer more educators access to NASA educational materials, NASA has formed partnerships with universities, museums, and other educational institutions to serve as Regional Teacher Resource Centers (RTRCs). For the location of the RTRC nearest you, please contact the TRC serving your geographic region.

NASA's Central Operation of Resources for Educators (CORE) was established as part of the Teacher Resource Center Network to facilitate the national and international distribution of NASA-produced educational materials in audiovisual format. Orders are processed for a small fee that includes the cost of the media. Send a written request on your school letterhead for a catalog and order forms. For more information, contact:

NASA CORE
Lorain County Joint Vocational School
15181 Route 58 South
Oberlin, OH 44074
(216) 774-1051, ext. 293 or 294

The following *Toys in Space* science kits are available through TEDCO, Inc.:

Item no.	Description	Price
44401	Kit 1—Introduction to Microgravity	$15.99
44402	Kit 2—Action/Reaction	$15.99
44403	Kit 3—Spinning Toys	$15.99
44404	Kit 4—Trajectory Toys	$15.99
00600	Gyros Gyroscope	$ 8.00
44444	Physics of Toys Video (50 minutes)	$16.99

Shipping and handling charges: Add $5 for orders up to $50 and $1.50 for each additional $10. Send check or money order to:

TEDCO, Inc./Dept. B
498 S. Washington St.
Hagerstown, IN 47346

If ordering by VISA or MasterCard, you may call (800) 654-6357 and ask for Department B.

Glossary of science terms

acceleration The rate of change in velocity.

action force A force exerted on an object.

air resistance Force of the air pushing against a moving object.

amplitude The distance that a moving wave rises or falls.

angular momentum A property of spinning motion that must be conserved. Angular momentum is the product of an object's mass, the radius of its circular path, and its velocity.

attitude The orientation or tilt of a spacecraft.

Bernoulli's principle A property of fluid flow indicating that when the velocity of an incompressible fluid increases, the pressure also increases.

buoyancy An upward force exerted on an object in a liquid equal to the weight of the liquid the object displaces. Zero gravity is a neutral buoyancy condition.

center of mass The point at which the entire mass of an object can be supported in a gravity field.

centrifugal force The apparent outward force exerted by an object moving in a circle. In reality, the object is simply trying to move in a straight line.

centripetal force The inward force that causes an object to turn.

compression Concentration of particles in a longitudinal wave.

crest The high point in a wave.

drag The resistance force exerted by a fluid when an object moves through it.

elastic and inelastic collisions For perfectly elastic collisions, the relative speed of recession after the collision equals the relative speed of approach before the collision. In a perfectly inelastic collision, no relative speed exists after the collision. The objects stay together.

elastic potential energy Term used to describe the energy stored in a stretched object (usually a spring).

energy The ability to do work (or to produce change).

force A push or pull.

freefall The condition of an object falling in a gravity field.

frequency The number of waves passing a point per unit time.

friction   A resistance force opposing motion.

g-force   The ratio produced when the force felt by an object is divided by the force that the object would feel while motionless on the earth's surface.

gravitational potential energy   Energy possessed by an object that can fall.

gravity   The force of attraction between masses.

gyroscopic stability   A term describing the resistance of a spinning object to any force that would change the orientation of the object's spin axis.

heat energy   The energy associated with moving molecules.

inertia   The property by which an object tends to remain at rest or in uniform straight-line motion.

kinetic energy   The energy possessed by an object because of its motion.

law of universal gravitation   All particles exert a gravitational force of attraction on one another. The magnitude of the force is directly proportional to the product of the masses of the particles and inversely proportional to the square of the distance between them.

longitudinal wave   A wave that vibrates back and forth in the direction of the wave's motion. Also called a compression wave.

Mach one   The speed of sound.

magnetism   A property of certain objects in which there is an attraction to unlike poles of other objects. Its origin lies in the orientation of atoms within the object.

mass   The amount of material an object contains.

microgravity   Almost weightless condition—only very small forces remain. This term is used by NASA to describe the on-orbit environment.

momentum   The product of an object's mass times its velocity. Momentum is a conserved quantity within a closed system.

Newton's laws   The three basic laws of motion first formulated by Sir Isaac Newton. (See chapter 2.)

node   A point in a standing wave where no motion occurs.

parabola   One possible path of an object falling freely in a gravity field. A tossed ball follows a parabolic arc.

photon   A packet of radiant energy.

potential energy   Energy required to place an object in a position. This energy is stored in the object until the object moves.

precession   The wobbling of a spinning object.

rarefaction   The part of a longitudinal wave where the density is lower than the surrounding medium.

reaction force The force exerted by an object experiencing an action force. The reaction force is equal to the action force, but in the opposite direction.

surface tension The strength of the boundary film at the surface of a liquid.

speed The distance traveled by an object divided by the time required to cover the distance.

torsional wave A wave caused by twisting a coiled spring.

transverse wave A wave in which vibrations are to the left and right as the wave moves forward.

trough A wave valley.

velocity The speed and direction of an object's motion.

wavelength The distance between two identical points in a wave (i.e., from crest to crest)

weight The force of the earth's gravity pull.

weightlessness Another name for freefall.

zero gravity Another name for freefall.

Index

Other Bestsellers of Related Interest

200 ILLUSTRATED SCIENCE EXPERIMENTS FOR CHILDREN—Robert J. Brown

An ideal sourcebook for parents, teachers, club and scout leaders, or anyone who's fascinated with the wonders of science, this outstanding book is designed to make learning basic scientific principles exciting and fun. Literally crammed with different and interesting things to keep your youngsters entertained for hours, the collection of experiments presented here demonstrates such principles as sound, vibrations, mechanics, electricity, and magnetism. 196 pages, 200 illustrations.

Paperback	157390-9	**$9.95**

ENVIRONMENTAL SCIENCE: 49 Science Fair Projects—Robert L. Bonnet and G. Daniel Keen

Here's a collection of fun educational projects that introduces children ages 8 through 13 to the effects of pollution, landfill decomposition, water contamination, chemical waste, and environmentally stressed wildlife. All projects are designed for use in science fair competitions and include an outline, a hypothesis, a complete materials list, and step-by-step procedures. 140 pages, illustrated.

Paperback	156104-8	**$9.95**
Hardcover	156094-7	**$17.95**

PHYSICS FOR KIDS: 49 Easy Experiments with Electricity and Magnetism—Robert W. Wood

What makes a magnet stick to the refrigerator? What makes the batteries in a flashlight work? Find the answers to these questions, and more, in this entertaining and instructional project book. These quick, safe, and inexpensive experiments include making items like: a magnet, potato battery, flashlight compass, telegraph, model railroad signal, and electric lock. 142 pages, 151 illustrations.

Paperback	156616-3	**$9.95**
Hardcover	156606-6	**$16.95**

SCIENCE FOR KIDS: 39 Easy Astronomy Experiments—Robert W. Wood

While learning about the wonders of the sky, kids will classify stars by temperature, use refracting and reflecting telescopes, photograph star tracks, and much more. 154 pages, Illustrated.

Paperback	156195-1	**$9.95**

Look for These and Other TAB Books at Your Local Bookstore

To Order Call Toll Free 1-800-822-8158
(24-hour telephone service available.)

or write to TAB Books, Blue Ridge Summit, PA 17294-0840.

Title	Product No.	Quantity	Price

☐ Check or money order made payable to TAB Books

Charge my ☐ VISA ☐ MasterCard ☐ American Express

Acct. No. _____ Exp. _____

Signature: _____

Name: _____

Address: _____

City: _____

State: _____ Zip: _____

Subtotal	$ _____
Postage and Handling ($3.00 in U.S., $5.00 outside U.S.)	$ _____
Add applicable state and local sales tax	$ _____
TOTAL	$ _____

TAB Books catalog free with purchase; otherwise send $1.00 in check or money order and receive $1.00 credit on your next purchase.

Orders outside U.S. must pay with international money in U.S. dollars drawn on a U.S. bank.

TAB Guarantee: If for any reason you are not satisfied with the book(s) you order, simply return it (them) within 15 days and receive a full refund.

BC

ABOUT THE AUTHOR

Dr. Carolyn Sumners designed the Toys in Space program at the Johnson Space Center. She is the director of astronomy and physics for the Houston Museum of Natural Science. She receives partial funding from the Houston Independent School District for her astronomy teaching and has taught more than three-quarter million students in her years at the museum. Dr. Sumners directs the museum's Burke Baker Planetarium, Challenger Learning Center, and rooftop Brown Observatory and astronomy lab. She is curator for the museum's hands-on Discovery Place and the new interactive Chemistry Hall. She also writes planetarium shows and conducts astronomy and computer programming courses for children, families, teachers, and astronauts.

Dr. Sumners is the principal investigator of the Science Connection, a $1.6 million elementary science education curriculum project being developed at the museum and funded by the National Science Foundation. She is codirector and principal writer of the Informal Science Study at the University of Houston. In the past five years she has conducted more than one hundred science teacher workshops at the regional and national level.

Dr. Sumners has just finished a book on dinosaurs and a new science discovery guide for Astroworld Theme Park.